THIRD EDITION

ADVANCED PRACTICAL ORGANIC CHEMISTRY

THIRD EDITION

ADVANCED PRACTICAL ORGANIC CHEMISTRY

JOHN LEONARD
BARRY LYGO
GARRY PROCTER

CRC Press
Taylor & Francis Group
Boca Raton   London   New York

CRC Press is an imprint of the
Taylor & Francis Group, an **informa** business

CRC Press
Taylor & Francis Group
6000 Broken Sound Parkway NW, Suite 300
Boca Raton, FL 33487-2742

Printed in the United States of America on acid-free paper
Version Date: 20121120

International Standard Book Number: 978-1-4398-6097-7 (Paperback)

Visit the Taylor & Francis Web site at
http://www.taylorandfrancis.com

and the CRC Press Web site at
http://www.crcpress.com

Contents

List of Figures

List of Tables

Preface

The preparation of organic compounds is central to many areas of scientific research, from the most applied to the most academic, and is not limited to chemists. Any research that uses new organic chemicals, or those that are not available commercially, will at some time require the synthesis of such compounds.

This highly practical book, covering the most up-to-date techniques commonly used in organic synthesis, is based on our experience of establishing research groups in synthetic organic chemistry and our association with some of the leading laboratories in the field. It is not claimed to be a comprehensive compilation of information to meet all possible needs and circumstances; rather, the intention has been to provide sufficient guidance to allow the researcher to carry out reactions under conditions that offer the highest chance of success.

The book is written for postgraduate and advanced level undergraduate organic chemists and for chemists in industry, particularly those involved in pharmaceutical, agrochemical, and other fine chemicals research. Biologists, biochemists, genetic engineers, material scientists, and polymer researchers in academia and industry will find the book a useful source of reference.

Authors

John Leonard is currently a principal scientist at AstraZeneca Pharmaceuticals, where he is primarily involved with synthetic route design and development activities. Prior to this, he was a professor of organic chemistry at the University of Salford (U.K.).

Garry Procter is a professor and director of teaching in the School of Chemistry at The University of Manchester (U.K.), and before this he was director of undergraduate laboratories in the Department of Chemistry and Chemical Biology at Harvard University.

Barry Lygo is currently professor of chemistry at the University of Nottingham (U.K.), working in the field of asymmetric catalysis and synthesis.

chapter one

General introduction

The preparation of organic compounds is central to many areas of scientific research, from the most applied to the most "academic," and is not limited to chemists. Any research that uses new organic chemicals, or those that are not available commercially, will at some time require the synthesis of such compounds. Accordingly, the biologist, biochemist, genetic engineer, materials scientist, and polymer researcher in a university or industry all might find themselves faced with the task of carrying out an organic preparation, along with those involved in pharmaceutical, agrochemical, and other fine chemicals research.

These scientists share with the new organic chemistry graduate student a need to be able to carry out modern organic synthesis with confidence and in such a way as to maximize the chance of success. The techniques, methods, and reagents used in organic synthesis are numerous and increasing every year. Many of these demand particular conditions and care at several stages of the process, and it is unrealistic to expect an undergraduate course to prepare the chemist for all the situations that might be met in research laboratories. The nonspecialist is even more likely not to be conversant with most modern techniques and reagents.

Nevertheless, it is perfectly possible for both the nonspecialist and the graduate student beginning research in organic chemistry to carry out such reactions with success, provided that the appropriate precautions are taken and the proper experimental protocol is observed.

Much of this is common sense, given knowledge of the properties of the reagents being used, as most general techniques are relatively straightforward. However, it is often very difficult for the beginner or nonspecialist to find the appropriate information.

All three of us were fortunate enough to gain our initial training in top synthetic organic research laboratories around the world and we have subsequently acquired over 80 years combined experience in the training and development of organic research chemists in industry and academia. The knowledge that we have gained over this time is gathered together in this book, in the hope that it will be an aid to the specialist and the nonspecialist alike. Of course, most research groups will have their own modifications and requirements, but on the whole, the basic principles will remain the same.

This book is intended to be a guide to carry out the types of reactions that are widely used in modern organic synthesis and is concerned with basic techniques. It is not intended to be a comprehensive survey of reagents and methods, but the appendices do contain some information on commonly used reagents.

If we have achieved our aims, users of this book should be able to approach their synthetic tasks with confidence. Organic synthesis is both exciting and satisfying and provides opportunity for real creativity. If our book helps anyone along this particular path, then our efforts will have been worthwhile.

chapter two

Safety

2.1 *Safety is your primary responsibility*

Chemical laboratories are potentially dangerous workplaces and accidents in the laboratory can have serious and tragic consequences. However, if you are aware of potential hazards and work with due care and attention to safety, the risk of accidents is small. Some general guidelines for safety in the laboratory are presented in this section. In addition to these principles, you *must* be familiar with the safety regulations in force in your area and the rules and guidelines applied by the administrators of your laboratory.

Your supervisor has a responsibility to warn you of the dangers associated with your work, and you should always consult him/her, or a safety officer, if you are unsure about potential hazards. However, your own safety, and that of your colleagues in the laboratory, is largely determined by *your* work practices. Always work carefully, use your common sense, and abide by the safety regulations.

Some important general principles of safe practice are summarized in the following rules:

1. Do your background reading and assessment of hazards *first*. Look for methods that involve the least hazardous reagents and techniques.
2. Assess all the possible hazards before carrying out a reaction. Pay particular attention to finding out about the dangers of handling unfamiliar chemicals, apparatuses, or procedures and make sure that any necessary precautions are in place before starting the experiment.
3. Work carefully—do not take risks. This covers basic rules such as always wearing safety spectacles and protective clothing, not working alone, and working neatly and unhurriedly.
4. Know the accident and emergency procedures. It is vital to know what to do in case of an accident. This includes being familiar with the firefighting and first aid equipment and knowing how to get assistance from qualified personnel.

2.2   Safe working practice

It has been emphasized already that you should be familiar with the regulations and codes of practice pertaining to your laboratory. All laboratories should work with fundamental safety principles, which form the "basis of safety," and you should make sure that you and the people that work around you are familiar with these. We will not discuss safety legislation here but some fundamental universal rules should be stressed. Never work alone in a laboratory, unless special safety arrangements have been put in place to comply with local regulations. Always wear suitable safety spectacles and an appropriate laboratory coat and use other protective equipment such as gloves, face masks, or safety shields if there is a particular hazard or local requirement. Never eat, drink, or smoke in a laboratory. Work at a safe, steady pace and keep your bench and your laboratory clean and tidy. Familiarity breeds contempt; do not allow yourself to get careless with everyday dangers such as solvent flammability. Familiarize yourself with the location and operation of the safety equipment in your laboratory.

As regards specific hazards, the chief rule is to carry out a full assessment of the dangers involved before using an unfamiliar chemical or piece of apparatus. Some of the most common hazards are described in Section 2.3. Once you are aware of the possible dangers, take all the necessary precautions and ensure that you know what to do if an accident or spillage occurs.

Store your chemicals in clearly labeled containers and abide by the regulations concerning storage of solvents and other hazardous materials. Dispose of waste chemicals safely, according to the approved procedures for your laboratory. Never pour organic compounds down the sink.

2.3   Safety risk assessments

Always assess the risks involved *before* carrying out a reaction. It is good practice to carry out a systematic risk assessment for any new experiment that you intend to carry out, reviewing the hazards associated with the chemicals being used as well as the equipment and experimental procedure. In most areas, safety legislation makes it mandatory to conduct such a safety audit, but even if it is not legally required, it should still be regarded as an essential preliminary before starting a reaction. If you are aware that a procedure might carry some risks, consider alternative ways of performing the experiment. Look for methods that involve the least hazardous reagents and techniques. Assess all the possible hazards associated with the reactions that you are planning to carry out as well as the chemicals you are using before carrying out a reaction. As well as toxicological issues, make sure that you are aware of any significant exotherms associated with the reaction and thermal instability issues associated with your chemicals. This is particularly important when you work on a larger scale. Remember that procedures other than chemical reactions can also be hazardous! Unstable compounds (see discussion in Section 2.4.3) can explode spontaneously when heated, for example, for the purpose of distillation or drying. Some of the worst personal accidents have been caused by applying mechanical energy to unstable compounds, for example, by grinding. Pay particular attention to finding out about the dangers of handling unfamiliar chemicals or apparatuses and make sure that any necessary precautions are in place before starting the experiment. Hazardous material and reactions can often be handled with complete safety when appropriate procedures are followed, but careful experimental design may be required.

2.4   Common hazards

Each experiment will have its own set of risks that should be taken into account, but a range of the more common general risks associated with carrying out chemical reactions are described in the following sections.

2.4.1   Injuries caused by use of laboratory equipment and apparatus

A high proportion of accidents in the laboratory occur when handling glassware. Hand injuries are perhaps the most common of injuries and can be serious. Many accidents occur when connecting rubber tubing to glassware, and inexperienced workers are particularly prone to such injuries, so learn from more experienced colleagues how to carry out such tasks safely. Stabbing injuries, when using Pasteur pipettes and syringes, are also common and can have dangerous consequences. Also pay attention to the condition of glassware and in particular check flasks for star cracks. Some general hazards are listed in Table 2.1, but there are also many other types of equipment in modern laboratories that have particular associated hazards.

2.4.2   Toxicological and other hazards caused by chemical exposure

Extensive compilations of information about the dangers posed by a large number of compounds are available (see Bibliography). Consult these references and your supervisor before using a compound or procedure that is new to you. Most chemicals are supplied with an extensive range of safety data, and these should examined before using any

Table 2.1 Common Hazards with Apparatus in the Chemical Laboratory

Source	Hazard
Electrical equipment	Danger of fires caused by electrical sparks with solvent vapors, and a risk of electrocution with badly maintained equipment
Glassware	Danger of cuts, particularly when handling glass equipment or when equipment is pressurized or evacuated. Faulty glassware can also lead to leaks of harmful compounds
Vacuum apparatus and glassware connected to it	May implode violently
Pressure apparatus and glassware connected to it	May explode violently
Gas cylinders	May leak harmful gases or discharge violently
Robotic equipment	Danger from unexpected movements

material that you are not familiar with. If you routinely analyze safety information, you will quickly become familiar with standardized hazard and precautionary (H/P) codes that are associated with particular compounds, and this makes it easier to assess any potential hazards quickly. Current information on chemical hazard risk and safety statements is easily found by searching the Internet, and a globally harmonized system (GHS) of chemical hazard labeling is due to be introduced by the UN in 2015. Remember to treat all compounds, especially new materials, with care. Avoid breathing vapors and do not allow solids or solutions to come into contact with your skin. The majority of accidents are caused by a few common hazards. Some frequently encountered dangers are listed in Table 2.2, and you should be aware of all of these and always take appropriate precautions. For some chemicals, the prime hazard is caused by acute corrosive effects to the skin, whereas others have toxic effects caused by ingestion. As well as avoiding any exposure to chemicals that have high toxicity levels, it is important to remember that long-term exposure to small amounts of less toxic materials (e.g., solvents) can also be dangerous.

2.4.3 *Chemical explosion and fire hazards*

A number of commonly encountered reagents are particularly hazardous because they are pyrophoric—they can spontaneously ignite when exposed to air or to moisture in the atmosphere. Common pyrophoric

Table 2.2 Common Chemical Exposure Hazards

Source	Hazard
Strong acids	Extremely corrosive. React violently with water, bases. May produce harmful vapors
Strong bases	Extremely corrosive. React violently with acids, protic solvents
Strong oxidizing agents	Extremely corrosive—cause skin burns
Alkali metals	React violently with water, protic solvents, and chlorinated solvents
Solvents	Solvent hazards are magnified because they are used in relatively large quantities. Most are highly flammable. Many are highly toxic, particularly, for example, halogenated solvents and some aromatics such as benzene
Alkylating agents, for example, methyl iodide and dimethyl sulfate	Extremely toxic—carcinogenic
Halides (fluorine, chlorine, and bromine)	Acute toxicity
Hydrofluoric acid and metal fluorides	Toxic and corrosive
Cyanides and hydrogen cyanide (HCN)	Acute toxicity
Oxalic acid and its salts, oxalyl chloride	Toxic to the kidneys
Aromatic amines and nitro compounds	Genotoxic and/or carcinogenic
Ozone	Harmful to respiratory system and eyes
Hydrogen sulfide	Acute toxicity, similar to HCN. Sense of smell is quickly deadened by exposure
Phosgene	Acute respiratory system toxicity
Osmium tetroxide	Extremely toxic at very low dose levels! Toxic effects include pulmonary edema and cornea damage/blindness
Benzene, polycyclic aromatics	Carcinogenic
Hexamethylphosphoric triamide (HMPA)	Carcinogenic

solids include finely divided reactive metals (e.g., Li, Na, and K), metal hydrides (e.g., NaH, KH, and $LiAlH_4$), and metal carbonyls (e.g., nickel carbonyl and dicobalt octacarbonyl). Transition metal catalysts, such as palladium and platinum on carbon and Raney nickel, can also ignite if not treated with care, particularly after use when they have hydrogen residues adsorbed on them. It is preferable to weigh these materials under inert atmosphere where possible. The less reactive materials can be weighed safely without an inert atmosphere, but sprinkling the fine powers through the air should be avoided. Metal alkyls, such as alkyllithium reagents and Grignard reagents, are also pyrophoric, but they are usually supplied in solution, which makes them safer to use. Note that even in solution, some of the more reactive reagents are pyrophoric. All concentrations of tert-butyllithium are pyrophoric, but it is less well recognized that concentrated solutions of n-butyllithim (e.g., 10 M) and s-butyllithium solutions are also pyrophoric (Table 2.3).

It is important to appreciate the reactive hazards of the material(s) to be handled either alone or in combination with other compounds. There are certain families of compounds that pose a severe risk of explosion, and some of these are listed in Table 2.4. The need to recognize the potential for unforeseen hazards such as violent exotherms or gas evolution is equally important. For this reason, it is wise to restrict the scale of a reaction the first time it is performed in the laboratory.

Consideration of the structure of compounds and reagents can provide an indication of severe reactivity, and the ability to recognize such potential hazards is part of being a good experimentalist. Several functional groups are particularly associated with the potential for a chemical to decompose violently or explosively, for example, nitro, acetylene, azide, peroxide, and peracid groups. Any compound containing these groups should be treated with utmost caution until proven safe. In general, compounds with structures that contain a high proportion of nitrogen and/or

Table 2.3 Common Pyrophoric Hazards

Alkali metals—for example, lithium, sodium, sodium potassium, and cesium

Metal hydrides and alkylated derivatives—for example, potassium hydride, sodium hydride, lithium aluminum hydride, diethylaluminum hydride, diisobutylaluminum hydride, and silanes

Finely divided metals—for example, bismuth, calcium, magnesium, titanium, and zirconium

Hydrogenation catalysts (especially used)—for example, Raney Ni, Pd/C, Pt/C

Alkylated metal halides—for example, dimethylaluminum chloride and titanium dichloride

Metal carbonyls—for example, dicobalt octacarbonyl and nickel tetracarbonyl

Reactive nonmetals—for example, white phosphorous

Solutions that should be treated with particular care:

Alkylated metals—for example, tert-butyllithium (all concentrations), n-butyllithium (10 M), and trimethylaluminum

Diborane (B_2H_6)—a gas that is normally used in solution

Table 2.4 Common Functional Groups with Chemical Explosion Hazards

---

Acetylene and metal acetylides
Azides ($-N = N^+ = N^-$), acyl azides ($-CO-N = N^+ = N^-$)
Metal azides ($M^+ N_3^-$)
Diazo compounds ($-C = N^+ = N^-$) and diazonium salts ($-N_2^+ X^-$)
Nitrates ($-ONO_2$)
Nitro ($-NO_2$) and polynitro compounds, for example, TNT (trinitrotoluene)
Nitrites ($-ONO$)
Nitroso ($-NO$)
Hydroxylamine derivatives ($-ONH_2$, $HO-NH-$, $-O-N-$)
Chlorates ($-ClO_3$), bromates ($-BrO_3$), iodates ($-IO_3$)
Perchlorates ($-ClO_4$)
Perchloric acid
Peroxides ($-O-O-$), hydroperoxides ($-O-OH$), peracids ($-CO_3H$)
Ozonides
Epoxides

Reactive mixtures

Liquid oxygen and liquid air (formed by evaporation of liquid nitrogen) in the
  presence of organic materials
Alkali metals in contact with chlorinated solvents
Oxidative groups listed above when mixed with reducing agents

---

oxygen atoms, relative to carbon atoms ([Number of C + N + O atoms]/
[Number of N + O] <3), tend to be unstable. This includes small organic
compounds containing one or more of the functional groups given in
Table 2.4.

Some common compounds are quite prone to the formation of peroxide derivatives in the presence of air or other sources of oxidation and
can accumulate simply through prolonged storage. Some of the most
commonly used "peroxidizable" compounds are ethereal solvents such as
diethyl ether, tetrahydrofuran (THF), dioxane, and diglyme. The presence
of peroxides is often overlooked, but they can be very dangerous and it is
therefore important to check ethers and other susceptible reagents for peroxide level before use. Even small amounts of peroxides in solvents can be
dangerous because they are generally nonvolatile and will accumulate in
the residues following distillations of large volumes of solvents (e.g., on a
rotary evaporator). Peroxides themselves can be explosive, and they can
also trigger other unexpected violent reactions with certain types of compound (e.g., peroxides may catalyze the violent polymerization of alkenes).
Compounds prone to peroxide formation are often supplied containing
peroxidation inhibitors, but these might be removed during subsequent
purification, rendering the material susceptible to peroxide formation. All
peroxidizable compounds/solvents should be stored in dark bottles away

from heat and prolonged storage should be avoided. Commercial indicator papers are available, which allow very quick testing for peroxides, and a supply of these should be kept in all organic chemistry laboratories.

Reactions that proceed with a significant exotherm are a common cause of laboratory hazard incidents, particularly when scaling up a reaction. Monitoring the internal temperature of reactions is a useful way of detecting sharply exothermic processes. With this knowledge, such reactions can often be carried out safely, even on a large scale, by good experimental design. A problem is commonly encountered when all the reagents are added at the start of the experiment—if a sharp exotherm occurs, the reaction can suddenly take off in an uncontrollable manner. This may lead to the reaction mixture being expelled from the reaction flask violently. Such problems can normally be avoided by adding a rate-limiting reagent slowly, maintaining the operating temperature of the reaction.

2.5   Accident and emergency procedures

Regrettably, accidents do sometimes happen, so it is vital that you know what to do if an accident does occur. You must be familiar with the fire-fighting equipment in your laboratory (fire extinguishers, fire blankets, and sand buckets) and you must know the procedures for summoning the fire department and for evacuating the building. In the case of injuries or exposure to harmful chemicals, you should know who to summon to administer first aid and how to get medical assistance. It is particularly important to know how to get help outside of normal working hours. If you are using particularly dangerous materials (such as cyanides) or equipment (such as high pressure apparatus), you should know about the relevant emergency procedures and take precautions such as having antidotes, protective equipment, and qualified personnel at hand. In the aftermath of an accident, it is very important that you complete the required accident report forms and take steps to avoid any possibility of a repeat. Ask yourself now: Are you familiar with accident procedures? If not you should *not* be working in the laboratory.

For more information on specific safety issues, consult safety manuals and other chapters of this book. Throughout the book, safety warnings are highlighted in ***bold italic*** text.

Bibliography

L. Bretherick, *Guide to Safe Practices in Chemical Laboratories*, Royal Society of Chemistry, London, 1986.

M.J. Lefevre, *First Aid Manual for Chemical Accidents*, Dowden, Hutchinson and Ross, Stroudsburg, 1980.

R.E. Lenga, Ed., *The Sigma-Aldrich Library of Chemical Safety Data*, 2nd ed., Sigma-Aldrich Corp., Milwaukee, WI, 1987.

G.G. Luxon, Ed., *Hazards in the Chemical Laboratory*, 5th ed., Royal Society of Chemistry, London, 1992.

D.A. Pipitone, *Safe Storage of Laboratory Chemicals*, Wiley, New York, 1984.

M.J. Pitt and E. Pitt, *Handbook of Laboratory Waste Disposal*, Wiley, New York, 1985.

N.I. Sax and R.J. Lewis, *Dangerous Properties of Industrial Materials*, 7th ed., Van Nostrand Reinhold Co., New York, 1988.

P. Urben, Ed., *Bretherick's Handbook of Reactive Chemical Hazards*, 7th ed., Academic Press, Oxford, U.K., 2006.

chapter three

Keeping records of laboratory work

3.1   Introduction

No matter how high the standard of experimental technique employed during a reaction, the results will be of little use unless an accurate record is kept of how that reaction was carried out and of the data obtained on the product(s). Individuals or individual research groups will develop their own style for recording experimental data, but no matter what format you choose to follow, there are certain pieces of vital information that should always be included. In this section, a format for keeping records of experimental data is suggested, and although this need not be strictly adhered to, it will be used to point out the essential information that needs to be included. It is suggested that records of experimental work and experimental data be kept in two complementary forms: the *laboratory notebook* should be a diary of experiments performed and should contain exact details of how experiments were carried out and a set of *data records* should also be kept to record the physical data and preferred experimental procedure for each individual compound that has been synthesized. Traditionally, both of these documents were either hard-backed notebooks or paper sheets in loose-leaf binders. At the time of writing this edition, most data records will be kept in some sort of electronic form, and many organizations have moved to electronic laboratory notebooks (ELNs) for recording experimental results. Whether paper or electronic records are kept, the data that you need to record and maintain is the same, so this chapter will concentrate on providing advice on the data that should be captured. Some advice will be provided regarding the different requirements for keeping electronic versus paper records.

3.2   The laboratory notebook

3.2.1   Why keep a laboratory book?

The form of your laboratory notebook will probably be defined by the organization or research group in which you work, but before any practical work is undertaken, you should be clear about the format you will use for making a record of your work. If the exact format for recording experiments is not prescribed, spend some time devising a layout for your laboratory notebook entries. It should be stressed that a laboratory book is not a format for polished report writing, but a daily log of work carried out in the laboratory.

Some of the main reasons for keeping a laboratory book are as follows:

1. It should record your scientific reasoning. It is most important that each experiment starts with a clear "Aim and Objective" and ends with a "Conclusion" that is related to the aim and objective of the experiment. It is good practice to summarize the conclusions from related experiments at appropriate intervals and use your combined conclusions to drive the aims of subsequent experiments—this is the "scientific method."
2. It should be a record of the exact procedure that was followed for the entire experiment. Even for reactions that do not proceed to give the desired product, it is important to have a good record of the procedure that can be referred to later. For instance, after several attempts to bring about a reaction have failed, it is often possible to learn from what has been done previously to design and carry out a more successful procedure.
3. It should be the main index point that will enable you to find experimental, literature, and spectroscopic data on any compound that you have synthesized.
4. It is the main source of reference when you write reports, papers, theses, and so on.
5. Importantly, it is the point of reference for other workers to follow your work, so it is very important that it is presented in a style that is legible and easy to understand.
6. It is a chronological diary of the experiments carried out, and thus, it should allow you to say exactly when a particular experiment was carried out.

3.2.2   Laboratory records, experimental validity, and intellectual property

Work undertaken in a research environment is normally directed toward learning about the unknown and making discoveries. Such discoveries are of potential value to the scientific community through publication in scientific journals. Where new materials or processes are invented, there may also be a commercial value through patenting. In either case, it can be of crucial importance to be able to demonstrate the validity of the experimental results and exactly when the experimental work was carried out. It is vital that laboratory records are dated as they are carried out and that appropriate supporting analytical evidence is collected. For a discovery to be patentable, it must be novel, useful, and inventive (i.e., not obvious to someone skilled in the prior art in that field) and capable of being reduced to practice. A successful laboratory experiment normally completes the requirement for patent submission. In Europe, the date of patent submission is the one that counts for one application to be granted ahead of another, but in the United States and some other countries, it is the date on which the idea was reduced to practice that counts. For academic research, it is just as important to be able to verify the authenticity of experimental work for papers that are submitted for publication. The faithfulness of work published in journals or theses is occasionally challenged. Fraudulent results have been reported from time to time and there can be disputes over originality or plagiarism, so having evidence of when an experiment was carried out is always of the utmost importance.

For paper notebook experimental records, every experiment should be dated at the start and signed and dated when finished. In many laboratories, countersignatures are required, where a colleague signs and dates the experiment, and this may add some value if the authenticity of the experiment is questioned. Dated spectra and other analytical data to complement the experiment also add weight to its validity.

When an ELN is used to record experiments, the records should be captured in a version controlled relational database that is maintained and validated for the capture of electronic records and signatures. When a document is saved in this type of database, the time and date are permanently associated with it and the version is also saved indelibly. Any versions subsequently saved are kept, in addition to previous ones, to provide an audit trail. Strong evidence for the point in time at which an experiment was carried out is provided by the time/date stamp on an ELN record from a relational database. The evidence will be further enhanced by associated electronic analytical data, such as nuclear magnetic resonance (NMR), mass spectrometry (MS), infrared (IR), and layer chromatography (LC) data, which have also been captured in separate, secure databases.

3.2.3 How to write a laboratory book: Paper or electronic

One of the most important aspects of keeping a laboratory book is that it is written at the bench as you perform the experiments. This can be a challenge when using an ELN and it is most important when introducing this technology that provision is made for a computer terminal, laptop, or other appropriate input device to be located at the laboratory bench. *It is extremely bad practice to keep rough notes about experiments and then transfer the details to a laboratory notebook later.* This can cause many problems, for instance, the original notes can be lost; even with the strongest will, the exact truth often becomes distorted in transferring information to the laboratory book and small facts that may at the time seem unimportant are left out; it is also very easy to forget to rewrite an experiment altogether, especially if the reaction failed, and this can lead to wasting much time later. It is more important that the laboratory book be an accurate record of the way an experiment was performed than for it to be in your neatest writing, although of course it should be legible.

An example of a format that is effective for general synthetic chemistry is outlined in Figure 3.1. This can be adjusted to personal needs, but the essential features, which are listed in the following, should be included in any format chosen.

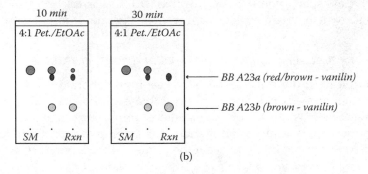

(a)

Ref., J.-L. Luche, L. Rodriguez-Hahn, P. Crabbe, J. Chem. Commun., 1978, 601.

Substance	Amount	Mol. wt.	M.moles	Equiv.	Source
BB A21a	200 mg	348	0.57		p. A21
NaBH₄	27 mg	38	0.71	2.84(H)	Aldrich
CeCl₃ (0.4 M in MeOH)	2 ml		0.8	1.4	
MeOH	25 ml				

Method:
The aldehyde (200 mg) and CeCl₃ solution (2 ml) in MeOH (25 ml) were cooled to 0°C, then NaBH₄ (27 mg) was added in batches (ca. 5 mg at a time) over 10 min.
TLC

(b)

After 30 min, TLC showed no starting material, so the MeOH was evaporated under reduced pressure. The residue was dissolved in CH₂Cl₂ (30 ml) and the solution washed with 10% aq. HCl (10 ml) followed by sat. aq. NaHCO₃ (3 × 10 ml), dried (MgSO₄) and then concentrated under reduced pressure. The crude product (210 mg) was isolated as a yellow oil.

Figure 3.1 An example of a laboratory notebook entry.

Flash chromatography using 9:1 (pet. ether/EtOAc) on 8 g of silica provided:

Rf 0.6: BB-A23a (27 mg, 12%)
^{1}H NMR (BB-A23a^{1}H)
^{13}C NMR (BB-A23a^{13}C)
IR (BB-A23aIR)
M/z (BB-A23aM/z)
looks like acetal:

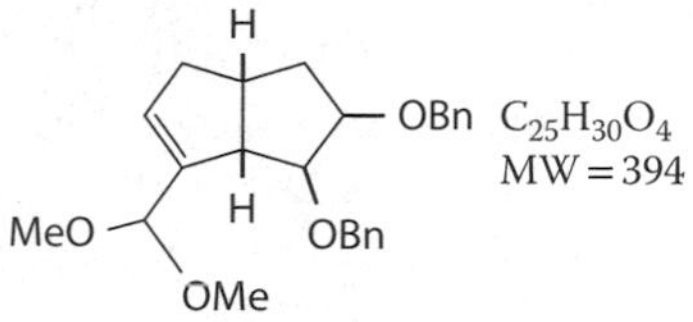

(c)

Rf 0.3: BB-A23b (140 mg (69%)
^{1}H NMR (BB-A23b^{1}H)
^{13}C NMR (BB-A23b^{13}C)
IR (BB-A23bIR)
M/z (BB-A23bM/z)
looks like desired alcohol:

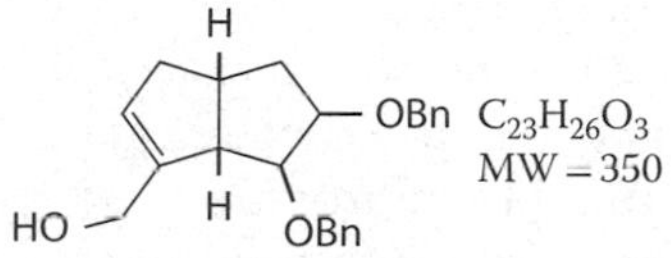

(d)

Comments:
Next time try aqueous solvent—may avoid acetal formation

Figure 3.1 An example of a laboratory notebook entry. (*Continued*)

3.2.4   *Paper laboratory notebook: Suggested laboratory notebook format*

1. *General layout*
   It is good practice to start each new experiment on the next free right-hand page of the laboratory book. This makes finding any particular experiment easier.
2. *Experiment number*
   Every experiment should be assigned a unique reference number. This number is usually placed in the top right-hand corner of the laboratory book page. This is very important since it is used to reference all the data generated from the experiment. If a laboratory book with numbered pages is used, it is common for the experiment number to include the number of the page on which the experiment starts. The way in which the laboratory book is indexed is open to personal preference, but usually includes the researcher's initials, the laboratory

book number, and the page number. Figure 3.1 illustrates an example of an experiment starting on page 23 of a researcher's first laboratory book (designated laboratory book A). Data and products from this experiment would then all be labeled with the number *A23*, prefixed with the researchers initials (in this case, *BB*). When more than one product is isolated from a reaction, a suffix, *a, b, c*, and so on, can be added to this reference number, *a* being the spot running highest on thin layer chromatography (TLC), *b* the next, and so on. Thus, for this experiment, two products were isolated, and these carry reference numbers *BB A23a* and *BB A23b*. Using this system, the origin of any synthetic sample and data set can be determined very quickly.

3. *The date*

The date should always be included as this serves as a record of exactly when the experiment was started.

4. *A reaction scheme indicating the proposed transformation*

This is always included at the top of the page so that an individual experiment is easily found. If the reaction proceeded as desired, the scheme is left intact, but if the desired product was not obtained, it can be crossed through in red to indicate this. If other products were also obtained, they can be added, again in a different color ink if desired. Thus, simply flicking through the laboratory book, looking at the schemes, can quickly provide a good deal of information. It is good practice to write a balanced equation that includes all by-products, both from the main reaction and from the reagents used. This can be helpful in working out mechanistic details or evaluating what happened if the reaction did not proceed as expected.

5. *Literature references*

It is good practice to write any key literature references in your laboratory book. These would typically be references to original reports of the procedure being followed or to publications containing characterization data for the compounds being prepared. Including this information in your laboratory book makes it very easy to find when writing reports and theses.

6. *Materials table of reagents and quantities*

A materials table should be included at the start of the write-up containing the quantities of each ingredient of the reaction, together with the molecular weight, number of moles, and number of molar equivalents. It is very useful to have this information available at a glance. Having the molecular weights in hand saves a good deal of time when going on to other reactions and when looking at mass spectra and so on, but the real importance of this section is that the values contained here are critical when evaluating the outcome of a reaction. These are the factors that you will want to vary in subsequent experiments when you need to optimize or change the outcome of the reaction in some way.

7. *The procedure*

This should be an exact account of the practical procedure carried out, including any spillages or other mishaps. It can be quite brief and not necessarily of publication standard, as long as it is understandable and accurate. It should include the temperature of the reaction and any observed changes, particularly temperature changes, that occur during the reaction or when reagents are added. Reaction procedures vary from simple to very complex, and all factors that may have an influence on the course of the reaction should be included, such as the type and speed of stirring, the method of cooling or heating, and the mode and rate of addition of reagents.

8. *Reaction monitoring*

TLC is still the most widely used method for reaction monitoring in academic laboratories, whereas in industry, high performance liquid chromatography (HPLC)/ultra performance liquid chromatography (UPLC) or liquid chromatography-mass spectrometry (LC-MS) are now commonly used, and it is likely that these techniques and/or direct MS techniques will become more universal in the future. When using TLC, it is very important to include a full-size representation of the TLC plate(s) in your laboratory book, giving the development solvent and the stain used for visualization. A series of TLC plates depicting the course of the reaction over time is worth many words when it comes to following a procedure. Some people find it convenient to draw the TLC plates on the adjacent laboratory book page (for more information on TLC, see Section 9.3.1).

The range of alternative reaction monitoring techniques that are available is expanding all the time and the one used will depend on the type of experiment. As well as chromatographic techniques, such as HPLC/LC-MS and gas chromatography (GC)/GC-MS, other commonly used profiling techniques include pH, temperature, turbidity, and in-line spectroscopic techniques such as IR, ultraviolet (UV), and NMR. An experimental write-up should contain reference to all the data that was used to draw the experimental conclusions, so it is preferable to attach a copy of any reaction monitoring that was used or provide a clear reference as to where it can be found.

9. *Details of work-up and purification of the product(s)*

If chromatography is used to purify the reaction product(s), it is important to include the quantity and type of adsorbent and the solvent system used for elution. Some people also like to include a TLC representation of the column fractions. If the product is purified by crystallization, record the solvent(s) used and the melting point. If it is distilled, describe the type of distillation setup and report the boiling point and pressure.

10. *Cross-references to the spectra and data files*

All compounds should be given unique reference numbers as described earlier, and these should be included on all spectra recorded on that compound. The yield of each compound isolated should also be given and if possible its structure. Where spectra and other analytical data are not attached to the notebook, their location (hard copy or electronic) should be provided.

11. *Finally, a clear conclusion to the experiment should be provided.*
    This might be something as simple as a statement that the reaction proceeded as expected. However, for a more complex study, it should be a summary of what was learned and how this learning will influence subsequent experiments.

3.2.5   *Electronic laboratory notebooks*

ELNs are a modern alternative to the traditional paper-based laboratory notebook discussed in Section 3.2.4. They are commonly used throughout the chemical industry and are increasingly being utilized in academic labs. ELNs are used in exactly the same way as traditional laboratory notebooks, but the exact format of the data entries will depend on the software used and the way the templates have been configured for your organization. If you are using an ELN that has been configured for synthetic chemistry, it will probably have a format similar to that described in Section 3.2.4. However, because ELNs are computer-based, they can offer a number of features that are not possible using traditional laboratory notebooks.

1. One ELN feature that is particularly attractive to users is the electronic materials table. This will have the capability of downloading or calculating properties of substrates and reagents, such as molecular weights, densities, molarities, and names. Once the amount of the limiting reagent and the required equivalents of other reagents are added, the quantities of reactants, including volumes of liquids, are calculated automatically. The yield of the reaction is also calculated once the product quantity is recorded.

2. Another feature of an ELN that removes a lot of tedium from writing up experiments is "cloning," where a completed experiment is used as the basis for a similar follow-up experiment.

3. Sophisticated search facilities allow you to rapidly locate and recover information based on structure, text, or data fields. This makes it important to think carefully about how you write up experiments so that you can gain the maximum benefit from searching. Newer ELN features also allow data to be compared across a number of experiments, which can be very useful.

4. The ability to instantly share information with coworkers and collaborators anywhere in the world.

5. The indelible time- and date-stamping of each stage of experiments is automatic, and updated versions are also saved permanently, but you must remember to save updates to the secure repository and it is also important to "publish" (sign-off) experiments that are complete.
6. The ability to automatically export experimental procedures and data to a report or thesis.
7. The ELN experiment also acts as a "paperclip" for renditions of key spectroscopic and analytical data, which are normally embedded as PDF files. However, it is good practice to keep the original raw data files (NMR flame ionization detectors [FIDs], etc.) in their dedicated data systems, with references to them in the ELN.

Whether you use a traditional laboratory notebook or an ELN will obviously depend on the local policy at your place of work. Either way, Section 3.2.3 will give you a good idea of the information that needs to be recorded.

3.3 Keeping records of data

When it comes to writing reports, papers, and especially theses, one of the most time-consuming and tedious jobs is collecting together experimental and spectroscopic data for compounds. It can be incredibly frustrating to find that a particular piece of data has been misplaced or was never obtained. Also, if data collecting is left until the time of report writing, errors can easily creep into spectral assignments. Consequently, it is essential to collect data and make spectral assignments as your work progresses and to keep this information stored in a standard format.

Whenever a significant compound has been synthesized, a data file should be created for it. Such data files are most useful if they are in electronic form, such that they can be reformatted to suit requirements for publication. As with experimental records, these files can be stored in an ELN or similar database to make them secure and searchable. The format of a data file can vary according to personal preferences, but it should contain at least the following pieces of information:

1. The structure and molecular formula of the compound.
2. An experimental procedure for the preparation of the compound, preferably in a style suitable for publication.
3. An appropriate range of spectroscopic and chromatographic data that is sufficient to characterize the compound. Full assignments of spectra should be entered in the data sheets as soon as the information is obtained, so when the time comes to write a report, most of the information required is directly on hand.
4. Cross-references to other spectra and the laboratory notebook.
5. Literature references, if there are any.

3.3.1 *Purity, structure determination, and characterization*

On preparing any compound for the first time, a competent organic chemist should always undertake a three-step procedure:

1. *Purification*
   The compound must be obtained in a high state of purity and be free of by-products and solvents.
2. *Structure determination*
   The structure of the compound must be established beyond any reasonable doubt, including stereochemistry, geometry, and so on.
3. *Characterization*

A range of data must be collected that will not only convince the wider chemical community of the structure and purity of the compound but will also serve as that compound's identity.

It is worth considering these three separate processes very carefully, so that whenever you prepare a compound you will always collect the appropriate data for it. The skills of purification, structure determination, and characterization are the mark of a competent organic chemist and you should strive to become as proficient as possible in each of them.

3.3.2 *What types of data should be collected?*

It should be clear from the preceding discussion that you need to collect a set of data on each compound you prepare to establish its purity and to determine its structure.

When you are determining the structure of a compound, it is essential to scrutinize each piece of evidence carefully and critically until you are convinced that all the data is in accord with your proposed structure. Sometimes, it will be quite easy to establish the structure of a compound beyond reasonable doubt, from, say, a simple ^{1}H NMR spectrum alone, especially if you know how the substance was synthesized. In other cases, structure determination may be a major project in itself. In particular, it is often a challenge to establish the absolute and/or relative stereochemical arrangement in chiral compounds. This may, for example, involve several sets of two-dimensional NMR correlation experiments to determine coupling constants and nuclear Overhauser effect (nOe) experiments to determine through-space interactions. It may be that a derivative of the original compound has to be made to obtain some crucial structural evidence. In some cases, you may even have to resort to x-ray analysis.

No matter how simple or lengthy the task of structure determination, you need to collect a full range of spectral data and some evidence of purity to convince the chemical community that you have characterized a "new" compound. The range of data required for any compound you synthesize will depend on several factors, but the most important of these are as follows:

1. Is the compound new or already known in the literature?
2. Where is the work to be reported?

Before you start to collect data for a compound, you should take these factors into account. For a known compound, you will need at least enough data to unambiguously establish that you have made the same material as is already published. If you have made a compound for the first time, you must characterize the compound fully, by collecting a comprehensive range of spectral data. The data must be in full accord with the proposed structure and be sufficient to convince any organic chemist that the correct assignment has been made. You must also have some clear evidence of purity.

Combustion analysis (within 0.3–0.5%) was traditionally used to determine the empirical formula of a compound and to establish that a compound is of "high" chemical purity. However, structural isomers cannot be distinguished by combustion analysis. Consequently, combustion analysis is not always sufficient and alternative purity tests combined with high-resolution MS (HRMS) measurements are now often required as proof of purity and molecular formula. Popular modern methods for purity analysis include ^{1}H NMR peak integration relative to an internal standard and HPLC or GC quantification relative to a sample of known purity.

The level of detail needed for compound characterization purposes will depend on the structural complexity of the molecule and on where the data will be reported. For the purpose of a patent application, a molecular mass measurement, evidence of purity, and a good set of NMR data that is in accord with the structure will normally be sufficient. If you aim to publish work in the chemical literature and/or thesis form, you should agree with your supervisor on the standard that is to be adopted. Each journal has its own specific requirements not only for the type of data required but also for the format in which it needs to be presented. These specifications are typically published on the journal publisher's website. Once you have decided which specification to comply with, make sure that you collect an adequate range of data for all your compounds as you prepare them. It will also save you many hours of work if you compile data in the format you have chosen from the outset of your work. For most purposes, the data specified as follows will normally be required for characterization purposes:

1. *Crystalline solid characteristics*
   For solids, the melting point should be specified, together with the solvent from which the compound was crystallized. If there is a distinctive crystal form and/or color, it should also be specified. A typical reporting format is colorless needles, m.p. 56°C–57°C (MeOH).

You should be aware that crystalline solids can often exist in a number of different polymorphic forms, depending on the solvent and conditions used for crystallization. Each of the polymorphs will have a different stability and the most stable one will have the highest melting point. x-ray powder diffraction (XRPD) is a simple x-ray scattering technique that provides a diffraction pattern that is characteristic of an individual polymorph and can be used to distinguish between them and/or detect polymorphic mixtures.

Differential scanning calorimetry (DSC) is an alternative to melting point determination and provides additional useful information. In DSC, the sample and a reference are heated in a calorimeter and the difference in the amount of heat needed to increase the temperature of the sample and reference is recorded. Melting is an endothermic process because heat is required to bring about the phase transition. The DSC experiment detects the melting range for the solid and also quantifies the heat required. Exothermic processes, such as conversion to a more stable polymorph or decomposition, are also detected and quantified, making DSC a useful safety measurement. An exothermic decomposition at low temperature is indicative of explosive behavior and indicates that caution is required when handling the compound.

2. *Boiling points of liquids*

For liquids that have been distilled, the boiling range and pressure should be specified. A typical reporting format is b.p. 120°C–122°C at 5 mmHg.

3. *Molecular formula determination*

HRMS is a commonly used technique for molecular formula determination. The observed mass for the molecular ion (or pseudomolecular ion) must normally be within 5 ppm of the calculated mass for electron impact (EI) measurements or within 10 ppm for chemical ionization (CI) and electrospray (ES) measurements. It is important to note that HRMS confirms that some molecules of a particular molecular formula are present in the sample but generally does not give any indication of purity, so some other evidence of compound purity will also be required. A typical reporting format for HRMS is Found $[M + H]^+$ 230.1393. $C_{11}H_{20}O_4N$ requires 230.1391.

Combustion analysis is the other main technique for molecular formula determination and has the advantage that it also provides an indication of sample purity. A typical format for reporting combustion analysis data is Found: C, 54.5; H, 5.8; N, 2.5%. $C_{25}H_{31}NO_{13}$ requires C, 54.2; H, 5.65; N, 2.5%.

4. *^1H NMR spectrum*

For organic compounds, ^1H NMR data is usually the most important piece of characterization data. You should try to analyze the spectrum as fully as possible, and the values for every individual proton chemical shift and every coupling constant should be identified if at all possible. Decoupling experiments and two-dimensional NMR experiments will often be required to do this. ^{1}H NMR data is often the most complex to report and so it is important that you record it in an appropriate format as soon as you have analyzed the spectrum. A high-field ^{1}H NMR spectrum can be used to provide evidence of purity and isomeric homogeneity, and most journals now require a copy of the spectrum as supplementary material to a publication.

A typical reporting format is as follows:

The nucleus should be specified as a subscript to the δ symbol, and the instrument frequency, solvent, and chemical shift standard should be given, for example, δ_H (300 MHz, CDCl$_3$).

Chemical shifts (in ppm) of individual signals and multiplets should be specified in sequence starting from the low δ end of the spectrum. How the spectrum was referenced is usually provided in a preamble.

Each chemical shift should be followed by a set of parentheses containing the following information, separated by commas, in the following order:

a. Number of nuclei under the signal, for example, 3 H.

b. Multiplicity—s, d, t, q, dd, and so on (br can be used for a broad signal).

c. Coupling constants, for example, $J_{1,2}$ 8, $J_{1,6}$ 3 given in Hertz (Hz) and listed in descending order.

d. An assignment of the proton, in the form CH$_3$CH$_2$ (italics are used to indicate which protons) or H-6 (the proton attached to C-6 in the structure).

Thus, a typical entry might begin as follows:

δ_H (300 MHz, CDCl$_3$) 1.84 (1H, ddd, $J_{6a,6b}$ 12.5, $J_{6a,7}$ 6.5, $J_{6a,5}$ 1.5, H-6a), 2.37 (1H, ddd, $J_{6a,6b}$ 12.5, $J_{6b,7}$ 11.0, $J_{6b,5}$ 9.0, H-6b), 2.68 (1H, m, H-5), 3.89 (4H, br. s, OCH$_2$CH$_2$O),

and so on.

5. *^{13}C NMR spectrum*

A fully characterized ^{13}C NMR spectrum is another essential piece of data for an organic compound and can also be used as evidence of purity. The number of H atoms attached to each carbon can be determined using a combination of DEPT-45, DEPT-90, and DEPT-135 experiments. DEPT-45 provides the entire ^{13}C NMR spectrum, DEPT-90 suppresses all signals except CHs, and DEPT-135 gives a spectrum in which the CH and CH$_3$ signals have the opposite phase to the CH$_2$ signals (C atoms with no protons attached are suppressed).

The resulting data can then be presented in the following form:

$$\delta_C \text{ (75 MHz, CDCl}_3\text{) 20.1 (CH}_3\text{), 44.6 (CH}_2\text{), 46.7 (CH),}$$

and so on.

6. *Additional NMR data*

    If other abundant NMR-active nuclei are present in the compound (e.g., ^{19}F, ^{31}P), then wherever possible, their chemical shifts should also be recorded and reported as part of the characterization of the molecule.

    There are also a variety of powerful modern NMR techniques that are often used to assist structure determination and spectroscopic assignment (e.g., nOe and two-dimensional correlation experiments). The data from such techniques should be reported as appropriate.

7. *IR spectrum*

    IR spectroscopy is generally no longer used as the primary means of characterizing a new organic compound, but it is still essential data. Do not forget to record the IR spectrum for *all* new compounds and record the frequencies of the main bands.

    This is typically presented in the following form:

$$\nu_{max}\text{(cm}^{-1}\text{) 2935 (CH), 1742 (C = O), 1200 (C-O).}$$

8. *Low-resolution mass spectrum*

    It is preferable to have a mass spectrum that shows the molecular ion of the compound, and soft ionization techniques such as ES, CI, and fast atom bombardment (FAB) are generally used for this purpose. EI ionization can also give molecular ion information, but often results in extensive fragmentation of the structure. As a consequence, EI tends only to be used when soft-ionization techniques fail or when fragmentation data is required to aid structure assignment.

    Whatever the method used, it is important to report the conditions under which the spectrum was run together with a list of the main peaks and their relative intensity. It is also useful to work out the structures of the fragment ions wherever possible.

    Low-resolution mass spectra are typically presented in the following form:

    M/z (CI, NH$_3$) 230 ([M + NH$_4$]$^+$, 100%), 213 ([M + H]$^+$, 11), 91 (50).

9. *Optical rotation data*

    If your compound is optically active, the specific rotation needs to be recorded. If it is a known compound, it is best to measure the rotation in the same solvent and at a similar concentration to that reported previously. If it is a new compound, then you should use a common inert organic solvent such as chloroform as the solvent and aim for a concentration of 0.7 g/dm^3. The solvent, concentration, and temperature should always be reported.

Optical rotation data is typically presented in the following form:

$$[\alpha]_D^{20} + 89 \,(c\; 0.7 \text{ in } CHCl_3)$$

Note that optical rotations are sometimes measured at frequencies other than that of the sodium-D line. If this is done, you will need to replace the "D" with the frequency used.

10. *Chromatography data*

HPLC and/or GC are often used to determine purity or isomer ratios present in a mixture. When using these techniques, it is essential to record the conditions, the means of detection, and the type of column used as well as retention times.

11. *Other data*

The preceding listed data are those that are essential for the characterization of most organic molecules. However, there are many other types of data that may need to be collected and reported (e.g., UV-vis, optical rotatory dispersion, and circular dichroism). Your supervisor should be able to advise on which additional types of data are appropriate.

3.3.3   Organizing your data records

It is essential that you organize your data records in a way that allows you to link a piece of data to the experiment that generated it and also allows you to rapidly locate all the data for a particular compound. There are various ways of doing this and your supervisor should be able to advise on the most appropriate method.

If you generate paper copies of data, we would suggest collating all the data for a given compound in a clear plastic folder. You can then label the folder with the experiment number (see discussion in Section 3.2.4) and compound structure and store them in an indexed ring file. It is also a good idea to include a summary data sheet in the front of each compound folder. Data sheets can be of a standardized design, with spaces for each type of data that is required. An example that would be suitable for a synthetic organic chemistry project is shown in Figure 3.2. The advantage of this system is that it is easy to see at a glance whether a particular piece of data has been obtained for a compound. However, two disadvantages of the fixed-format data sheet are as follows: it does not provide the flexibility that is often required to record diverse types of data used to characterize any particular compound, and it tends to encourage the mistaken idea that every compound requires the same characterization data.

An alternative is to have a more flexible format as shown in Figure 3.3. This would need to be generated within a word-processing package so that it can be easily modified. Then, when a new compound is synthesized, the

Data Sheet

Chem. Name					
Lit. Refs.					
Scheme					
Method					

Notebook Ref			Spectra Refs.		
Mol. Formula			m.p./b.p.		
Tlc - R_f (solv)			$[\alpha]_D$ (c, solv.)		
v_{max}/cm^{-1}					

^{1}H NMR	*δ value*	*No. H*	*Mult.*	*J-value/Hz (coupled proton)*	*Proton*

^{13}C NMR	
M/z	
HMRS	
Microanalysis	
Other data	

Figure 3.2 An example of a fixed-format data sheet.

data record can be tailored to the needs of that particular compound. Any of the data sections that are inappropriate for that particular compound can be deleted, and any additional data sections can be added. An additional advantage of generating your data sheets using a word processor is that the data can be entered in the format required for publication in reports, papers, or theses (and can easily be adjusted to other formats). Consequently, you never need to type out the data again. An example of a completed datasheet is shown in Figure 3.4 (see Section 3.4.3 for details of how this type of data record can be transformed into an experimental section entry).

If you use an electronic laboratory book (Section 3.2.5), it will almost certainly have the facility to embed analytical data within the experiment entry. This useful feature enables you to use the ELN rather like a folder for printouts of spectra and other data, although it should not become a raw data repository for unprocessed data such as NMR FIDs. You may be able to create a data sheet template as an ELN document or you can simply store files in a convenient electronic format. Either way, the principles

DATA SHEET

Chemical Name:

Scheme:

Lit. Refs:

Method:

Mol. Formula:
Notebook ref.:
Spectra refs.:
m.p./b.p.   °C
Tlc: R_f (/   )
[α]$_D$:   (c =   ,   )
V$_{max}$/cm^{-1}·
δ$_H$ (MHz,   ):
δ$_C$ (MHz,   ):
λ$_{max}$/nm:
m/z (   ):
HMRS: Found: M$^+$ C H N O   requires
Microanalysis: Found; C   %; H   %; N   %
                                C H N O   requires C   %; H   %; N   %

Figure 3.3 A flexible format data sheet (word processor file).

involved are exactly the same as described in Section 3.2.4 and the data you collect will be the same. One advantage of storing data sheets within an ELN is that you will probably be able to search for them by structure. In addition, the software may allow you to generate experimental reports directly from the entry. These features can save a lot of time, but it is important that you familiarize yourself with software and understand any limitations in the functionality at the start of your project to avoid problems later.

3.4   Some tips on report and thesis preparation

Most research projects will culminate in the submission of a report or thesis and the preparation of such documents can be quite daunting if you are inexperienced in such matters. In this section, we will try to provide some general guidance, paying particular attention to presentation

(±)-*exo, exo*-7,8-Dibenzyloxy-*exo-cis*-bicyclo[3.3.0]oct-2-en-4-ol-2-carboxaldehyde

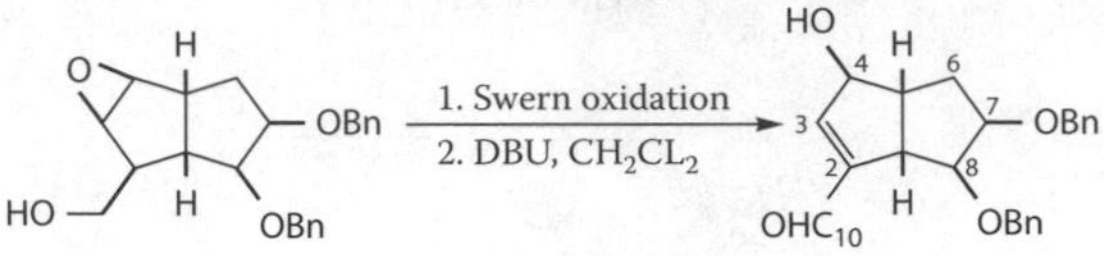

Method:

A stirred solution of oxalyl chloride (91 cm³, 1.0 mmol) in dry dichloromethane (15 cm³) was placed under a nitrogen atmosphere and cooled to −78°C. A mixture of dimethylsulfoxide (182 cm³, 2.35 mmol) and dichloromethane (1 cm³) was then added dropwise. After 5 min exo,exo-7,8-dibenzyloxy-exo-3,4-epoxy-exo-2-(hydroxymethyl)-cis-bicyclo[3.3.0]octane (200 mg, 0.55 mmol) in dichloromethane (2 cm³) was added dropwise and after a further 20 mintriethylamine (1 cm³, 7.61 mmol) was added. After 10 min at −78°C, the mixture was allowed to warm to room temperature, then partitioned between 2 M hydrochloric acid (30 cm³) and dichloromethane (2 × 30 cm³). The organic phase was washed with sat. aq. sodium hydrogen carbonate (40 cm³), dried (MgSO₄) and then concentrated under reduced pressure. The residue was dissolved in dichloromethane (15 cm³) and 1,5-diazabicyclo[5.4.0]undec-5-ene (167 mg, 1.1 mmol) added. The mixture was stirred at room temperature for 2 h, then poured into 2M hydrochloric acid (20 cm³) and extracted with dichloromethane (2 × 30 cm³). The combined extracts were washed with sat. aq. sodium hydrogen carbonate (40 cm³), dried (MgSO₄) and then concentrated under reduced pressure. Purification by flash chromatography (petroleum ether/ethyl acetate, 1/1) provided the title compound (159 mg, 80%) as a colorless oil.

Mol. Formula:	$C_{23}H_{24}O_4$ (MW = 364)				
Labook Ref:	BB-D45				
TLC:	R_f 0.3 (uv-active; pet. ether/ethyl acetate,1/1)				
v_{max}/cm^{-1}	3400 (OH), 3075, 3050, 2750 (CHO),1690 (C = O)				
δ_H (300 MHz, CDCl₃):	δ value	No.H	Mult	J value/Hz (coupled proton)	Proton(s)
	1.84	1	ddd	J_{6a6b}12.5, $J_{6a,7}$ 6.5, $J_{6a,5}$ 1.5	H-6a
	2.37	1	ddd	$J_{6b,6a}$ 12.5, $J_{6b,7}$ 11.0, $J_{6b,5}$ 9.0	H-6b
	2.68	1	m	$J_{5,6b}$ 9.0, $J_{5,1}$ 8.0, $J_{5,6a}$ 1.5	H-5
	3.47	1	ddd	$J_{7,6b}$ 11.0, $J_{7,6a}$ 6.5, $J_{7,8}$ 4.5	H-7
	3.58	1	ddd	$J_{1,5}$ 8.0, $J_{1,3}$ 1.0, $J_{1,8}$ 1.0	H-1
	3.80	1	dd	$J_{8,7}$ 4.5, $J_{8,1}$ 1.0	H-8
	4.28	1	d	J 11.0	CHaHbPh
	4.38	1	d	J 12.0	CHaHbPh
	4.50	1	dd	$J_{4,5}$ 1.5, $J_{4,3}$ 1.0	H-4
	4.68	1	d	J 12.0	CHaHbPh
	4.77	1	d	J 11.0	CHaHbPh
	6.55	1	dd	$J_{3,1}$ 1.0, $J_{3,4}$ 1.0	H-3
	7.11-7.50	10	m		Ar-H
	9.73	1	s		CHO
m/z (+ve FAB, thioglycerol):	365 ([M + H]⁺, 10%), 364 (M⁺, 7), 57 (100)				
Found:	[M + H]⁺ 365.1751. $C_{23}H_{24}O_4$ requires 365.1748				

Figure 3.4 A completed data sheet.

of experimental results. The backbone of any thesis or report in organic chemistry is the body of experimental results that have been gathered during the project. For this reason, the results should be reviewed and evaluated before starting to write the report.

3.4.1 Sections of a report or thesis

Most detailed organic chemistry reports will consist of three main sections:

1. *The Introduction*
   In this section, the project is introduced, the origins of the project and the original objectives are outlined, and all relevant background work is reviewed and referenced.
2. *The Results and Discussion*
   This is a detailed description of the work that was actually carried out. It should be a guided tour of the project presenting the reader with a realistic picture of how the project developed and the discoveries that were made during the course of the work. There should be discussions of objectives that were concluded successfully and explanations of those that failed.
3. *The Experimental Section*
   This section contains the method(s) of preparation for each compound, together with a set of data that is adequate to characterize it. The range of data presented and the style of presentation must conform to strict technical requirements, and it is most important that the appropriate standards are understood and adhered to.

3.4.2 Planning a report or thesis

Planning is the key to writing a good report or thesis. From the outset, you should aim to construct the document so that it has a logical structure and is easy to follow. The mark of a good report is that it will be a useful document for someone *less* knowledgeable than yourself, who may, for example, take over the project. A thesis that only your supervisor can understand will be of little use since he or she is presumably familiar with the results already. Always remember when you write a report that *you are the expert* and your aim is to produce a document that will be a useful resource for other workers.

If you have kept good data records for your compounds, as discussed in Section 3.3, you will have a good basis for starting your thesis plan. First of all, carefully review the original objectives of the project and then organize your compound data sheets into a logical order based on how the project progressed with respect to the objectives. This collection of data sheets will eventually become your experimental section.

Having reviewed your experimental work, you should have a clear picture of the key achievements made during the course of the project. During many research projects, the original objectives will not have been accomplished, and it may at first seem that much of the work has been unsuccessful. Nevertheless, it is most unusual if significant discoveries have not been made. You should construct your report so that achievements and discoveries are highlighted. Try to take an overview of your work so that you can partition it into separate topics if necessary. A large body of work, such as that contained in a thesis, is normally best broken down into sections or chapters of related work. Try to be logical about pieces of work that are grouped into chapters and sections and then present them in logical sequences. For example, a target synthesis project could be broken down into chapters describing different approaches to the molecule or chapters describing the synthesis of different fragments of the molecule. Try to find the best way of constructing the document so that a reader can follow a logical path through your results. Also, try to divide up the report so that a reader can easily find and interpret any individual topic. Once you have assessed your work in this way, you should have a working structure plan for your report. Do not treat this as a final plan and do not be afraid to rearrange sections or chapters as the report takes form.

Having reviewed your own work, you should be in a good position to appreciate what background material should be covered in the introduction. The nature of the introduction will depend on the type of project. For example, if the aim of the project was to synthesize a natural product, it would be appropriate to report the origin of the compound, its structure determination, properties, biosynthesis (if known), and then review any other synthetic approaches that have been reported. Your own synthetic strategy could then be explained as a lead-in to your discussion section. By the time you are ready to write a thesis or report, you should be familiar with the background to the project and you will probably have a large volume of background reference material. Before you start any further serious literature searching, make sure you are familiar with any previous reports relating to the project. It is a good idea to discuss the background to the project with your supervisor. Then, you will have to get down to some serious library work, searching for any relevant papers and reviews that you have not already read. See Chapter 17 for more information on literature searching.

The next stage is to create an outline plan for your review. Decide on the background material you wish to cover and group together papers that are related to one another. Groups of related papers will form the sections of your review. When arranging the material into a sequence, it is good practice to start from a broad base of background work and gradually focus on the material that is of direct relevance to your project. Try to order the material logically. Think back to when you started work on the project and try to construct the type of document that you would have ideally liked to have been given at that time. Once you have grouped together related pieces of background material and arranged the topics into a logical order, you will have an outline plan to start writing from.

3.4.3   *Writing the report or thesis*

Once you have a good thesis plan, you should be in a position to start writing effectively. Remember to break down the material into sections, the sections into subsections, and if necessary the subsections into sub-subsections, to make it easy for the reader to find particular pieces of information. Remember that very few people will ever want to read a scientific report from cover to cover; they will usually want to look up a particular section, so try to make them self-contained. The hierarchical sectioning system used in this book is now widely used and can be recommended.

Before you start writing, you should have copies of background reports and papers grouped together. It is a good idea to keep each group in a separate folder. Make sure you are fully conversant with each piece of work before you write about it and, when you are reviewing the work of others, be careful that you put the results in their appropriate context. Now that you have the material in manageable packages, you can start writing.

Start with the broader issues first, outlining the general area of chemistry into which the project belongs and then gradually focus on the material that directly relates to the project. Prepare reaction schemes that clearly illustrate each piece of work that you review. Good schemes are generally the most important aspect of a review. Be as concise as possible with the text, using it primarily to explain the information contained in the schemes. Describe the objectives of studies that have been carried out previously and summarize the results, but do not give lengthy details of experimental work.

As each background document is covered, add it to a reference list *in the proper format* with all the author names and initials, the year, volume, page range, and correct abbreviated journal name. It will save you so much time if you get this information detailed correctly from the outset. The format used for presenting references must be consistent and should conform to a recognized style (as required by a journal or your institute).

When you are writing the document, you will need to assign unique numbers to all the references, schemes, and structures to identify them in the discussion. Clearly, the numbering used in the first draft will almost certainly need to change as you add/remove components in later revisions. However, do not be tempted to write "XX" wherever a number is required. This does not save time and it makes the document impossible for a proofreader to understand. For example, "see Scheme XX" is meaningless if all schemes have been labeled "XX." Wherever possible, you should use the numbering that you intend to employ in the final document and then edit as required in subsequent revisions. This may not be possible if you are writing later sections of the document first. In this instance, you may have to assign unique numbers to the structures and references within the sections and then reedit the numbering system once the document is complete.

As you cover the topics, you will most probably want to arrange some sections in a different order than in your original plan and you may also decide that some extra ground has to be covered and/or that some of the originally planned material can be cut. Before you worry too much about the final ordering, try to write all the topics and make sure each is covered accurately. At this point, most of the technical work will have been done, but a good deal of work will still be required before the review is readable. You will need to organize the topics into their final arrangement within sections, subsections, and so on. Then, make sure that there is a flow through the document from one section to the next.

At this stage, your introduction may be quite bland and read like a list of reactions. There are several simple things you can do to make it more interesting for the reader:

1. Make sure each new topic is introduced so that the subject matter is put into some sort of context.
2. Compare and contrast results from different research groups.
3. Highlight what you consider to be the major achievements in the field before your study and explain why you think they are important.
4. Try to set the stage for your own study.
5. Finally outline the objectives of your work clearly and with respect to previous studies.

Once you are happy with the content and coverage of your introduction, let someone else read it, preferably your supervisor. This other person may be quite critical of your first draft, but do not be put off, in fact you should welcome constructive suggestions. If you have time, it is best to put the introduction aside for a few days before redrafting and editing. This will make it much easier for you to spot mistakes. When it comes to the final edit, take particular note of the following:

1. Make sure you have not left out any important points.
2. Make sure there is a logical structure and a good flow from one topic to the next.
3. Make sure the topics are balanced, trimming any that are too long and enhancing any that are too brief.
4. Do not be afraid to edit material that is not relevant and make all your writing as concise as possible.
5. Make sure you are satisfied that you have provided a good introduction to your project for anyone new to the field of study—this is the crucial test for an introduction!

When you have made all the final changes, you can then start at the beginning of the document numbering all the schemes, compounds, and references in sequence. Then, read and check through the document several times carefully, correcting mistakes and making final minor changes.

Most of the preceding advice also applies to the discussion section, but certain distinct differences in style may be required when reporting your own results. Once you have an outline plan, you can start writing your results within sections. Give your writing some structure and avoid sections that are simply repetitive lists of the experiments carried out. The following sequence is a useful way to describe work carried out within a topic or for describing an individual experiment:

1. Always start by outlining the objectives of the topic or the particular experiment that you are about to describe.
2. Outline the work or experiment that was carried out, pointing out any special features of the experiment(s). Note that when describing your own work, you will need to provide more detail than for the experiments described in your introduction.
3. Report the outcome of the experiment(s) and how you determined the outcome. This may be simple or it may involve a detailed analysis of complex analytical data.
4. Avoid bland statements such as "the reaction failed" or "the reaction gave 50% of the required product." Wherever possible, try to explain the fate of all the material that went into a reaction.
5. Draw conclusions from the outcome of the experiments and compare this to the original objectives.
6. Explain how the outcome led to follow-up studies.

This is basically a cyclic six-step process, where step 4 leads on to step 1 of the next experiment or topic. There will almost certainly be significant portions of your work that did not proceed according to the original objectives. However, it is rare that there is nothing to report concerning these "failed experiments." Indeed, it is often necessary to discuss the actual outcome of such reactions in some detail before proceeding to introduce alternative strategies that were adopted to solve the problem. Occasionally, results from a reaction that did not go according to plan are more interesting than the intended outcome. In such cases, a good deal of discussion will be required. Pay attention to the following points throughout the discussion:

1. Use reaction schemes to illustrate your work and make sure that they are located close to the text that refers to them. Avoid making the reader search for a numbered structure many pages from where it is described in the text; if necessary repeat the structure.
2. Use tables to compare results from related studies.

3. Try to present your studies in a positive manner.
4. Make connections between the work covered under different sections.
5. Where you have interesting results to report, highlight them and explain clearly the significance of your discoveries.
6. Compare and contrast your results with other published work, especially contemporary work.
7. Modulate your writing style and do not be afraid to point out when results are interesting, curious, confusing, incompatible with those in the literature, and so on.

The final stages of editing and redrafting the discussion are as described for the introduction earlier.

Unlike the other parts of the report, a bland, repetitive presentation style is an asset in the experimental section. Therefore, very little imagination is required here, but it is technically more demanding than the other sections. This section of a thesis will be measured against rigid standards and a badly prepared experimental section is likely to result in the thesis being failed.

The following guidelines provide some tips on how to write an experimental section that should be acceptable to a thesis examiner or journal referee:

1. Make sure you know what data is required *before your first day* in the laboratory. Start collecting good sets of data for all your compounds from day 1 and make sure you maintain this throughout your studies.
2. When writing your experimental section, be absolutely consistent about the style in which your data is presented and stick to an agreed journal or institutional standard.
3. Make sure you have a comprehensive selection of data for all new compounds, including proof of molecular formula and purity. Do not forget individual pieces of data that are easily overlooked, such as melting points and the specific rotation for optically active compounds.
4. Present your data to an accuracy within the limits of experimental error *under the conditions collected*, not according to the nominal tolerance limits of the instrument or as given in an electronic printout! For example, δ values from routine high-field ^{1}H NMR spectra should not be quoted to an accuracy of more than 0.01 ppm, and coupling constants should not normally be quoted to an accuracy of more than 0.5 Hz. ^{13}C NMR δ values should not be quoted to more than one decimal place, positions of IR bands should be given in whole wave numbers, melting points and boiling points should be

given to the nearest degree, and so on. Apart from anything else, data becomes less easy to decipher and is therefore *less* useful when quoted to higher tolerance limits.

If you follow these recommendations, preparation of your experimental section will be straightforward and you will avoid the problems encountered by many graduate students.

At the beginning of the experimental section, there should be a preamble describing the instrumentation used to record the data and the conditions that the data was recorded under. Be careful to check the make and model numbers of instruments and be sure you record conditions accurately. It is also useful to state the standard units used in the reported data, as this will save you repeating these units in each data set. It is also useful to record specifications for standard methods used, for example, the type of silica and TLC plates used for chromatography, the drying agent used for routine drying of organic solutions, and so on. Again, this will save a lot of duplication in each entry. An example of a typical experimental preamble is shown as follows:

"Melting point determinations were carried out on a Koffler block electrothermal apparatus and were recorded uncorrected. Infrared absorption spectra were run either neat (for liquids) or as Nujol mulls (for solids) on a Perkin-Elmer 1710 FT-IR instrument. ^{1}H NMR spectra were recorded at 300 MHz on a Bruker AC-300 instrument, as solutions in deuterochloroform, unless stated otherwise. Chemical shifts are referenced to tetramethylsilane and J values are rounded to the nearest 0.5 Hz. Mass spectra were recorded at low resolution on a Waters SQD Quadrupole instrument and at high resolution on a Waters Xevo G2 TOF instrument, under electrospray (ESI), electron impact (EI), or atmospheric pressure chemical ionization (APCI) conditions, as specified. After aqueous work-up of reaction mixtures, organic solutions were routinely dried with anhydrous magnesium sulfate. Thin layer chromatography was carried out using Merck Kieselgel 60 F_{254} glass-backed plates. The plates were visualized by the use of a UV lamp, or by dipping in a solution of vanillin in ethanolic sulfuric acid, followed by heating. Silica gel 60 (particle sizes 40–63 μm) supplied by E.M. Merck was employed for flash chromatography."

For preparations of organic compounds, the title of each data entry should usually be the systematic IUPAC[1] name of the compound being prepared, although a generic name sometimes is acceptable. Systematic names can be determined using a structure-to-name conversion tool, for example, as found in modern chemical drawing software. The name should be followed by a description of the method by which the compound was prepared. A more or less standard format has to be developed for reporting such methods. The method must be precise and concise. Repetitious statements should be avoided. For example, once it has been stated that the reaction mixture was stirred, or under nitrogen, or at −78°C, these facts need not be restated unless the conditions are altered. This is the one part of any report where all the individual write-ups should be presented in the same style. This is why it is worth investing time in learning the standard style at the very start of the project.

Suggested key phrases for a standard experimental method are given in Table 3.1.

Simply by changing the italicized text, this standard format can be used to describe a wide range of different reaction processes. It should immediately be followed by a list of characterization data outlined in Section 3.3.2. Figure 3.4 illustrates how the final entry might look.

If this experiment is included in word-processed data sheets (Section 3.3.3), experimental entries for reports and theses can be created easily. Simply by deleting headings, the data sheet shown in Figure 3.4 is transformed into a tabulated experimental entry, as shown in Figure 3.5. Alternatively, slight reformatting can also be employed to generate a journal-specific[2] experimental entry as shown in Figure 3.6.

Table 3.1 Suggested Key Phrases for a Standard Experimental Method

[Compound name] (A g, B mmol) was dissolved in [solvent name] (D cm³), the solution was then magnetically stirred under an atmosphere of nitrogen, and cooled to C°C. [Reagent name] (E g, F mmol) was added dropwise over G min	*This part describes the initial reaction conditions and addition of reagent(s) to substrates—it is essential to include the amounts in both g and mmoles*
After H h [aqueous solution name] (I cm³) was added, the organic layer was separated and washed with [aqueous solution name] (2 × J cm³), then dried over magnesium sulfate, filtered, and concentrated under reduced pressure	*This part says how long the reaction was left, how it was worked up, and how the crude product was isolated*
The [color, solid/liquid] residue was purified by [flash chromatography (solvent system)/distillation, b.p. K°C at L mmHg/recrystallization (solvent), m.p. M°C], to provide the title compound as a [color, solid/liquid] (N g, P%)	*This describes how the product was purified (choose the appropriate statement) and gives the yield of the purified product*

(±)-*exo*, *exo*-7,8-Dibenzyloxy-*exo*-*cis*-bicyclo[3.3.0]oct-2-en-4-ol-2-carboxaldehyde

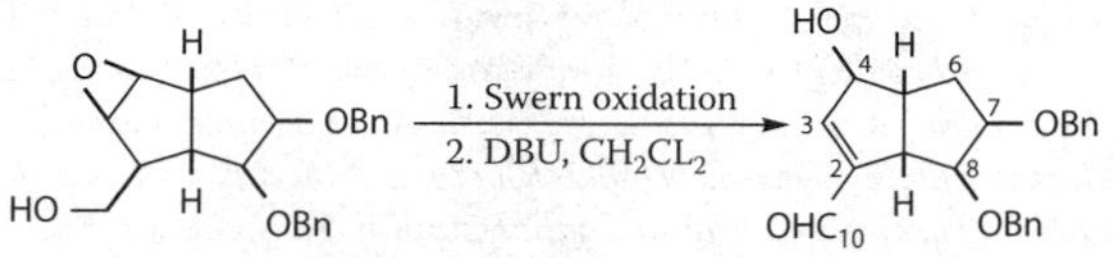

1. Swern oxidation
2. DBU, CH$_2$CL$_2$

A stirred solution of oxalyl chloride (91 cm^3, 1.0 mmol) in dry dichloromethane (15 cm^3) was placed under a nitrogen atmosphere and cooled to −78°C. A mixture of dimethylsulfoxide (182 cm^3, 2.35 mmol) and dichloromethane (1 cm^3) was then added dropwise. After 5 min exo,exo-7,8-dibenzyloxy-exo-3,4-epoxy-exo-2-(hydroxymethyl)-cis-bicyclo[3.3.0]octane (200 mg, 0.55 mmol) in dichloromethane (2 cm^3) was added dropwise and after a further 20 min triethylamine (1 cm^3, 7.61 mmol) was added. After 10 min at −78°C, the mixture was allowed to warm to room temperature, then partitioned between 2 M hydrochloric acid (30 cm^3) and dichloromethane (2 × 30 cm^3). The organic phase was washed with sat. aq. sodium hydrogen carbonate (40 cm^3), dried (MgSO$_4$) and then concentrated under reduced pressure. The residue was dissolved in dichloromethane (15 cm^3) and 1,5-diazabicyclo[5.4.0]undec-5-ene (167 mg, 1.1 mmol) added. The mixture was stirred at room temperature for 2 h, then poured into 2 M hydrochloric acid (20 cm^3) and extracted with dichloromethane (2 × 30 cm^3). The combined extracts were washed with sat. aq. sodium hydrogen carbonate (40 cm^3), dried (MgSO$_4$) and then concentrated under reduced pressure. Purification by flash chromatography (petroleum ether/ethyl acetate, 1/1) provided the title compound (159 mg, 80%) as a colorless oil.

TLC:	R$_f$ 0.3 (uv-active; pet. ether/ethyl acetate, 1/1)				
ν_{max}/cm^{-1}	3400 (OH), 3075, 3050, 2750 (CHO), 1690 (C = O)				
δ_H (300 MHz, CDCl$_3$):	δ value	no. H	Mult.	J value/Hz (coupled proton)	proton(s)
	1.04	1	ddd	$J_{6a,6b}$ 12.5, $J_{6a,7}$ 6.5, $J_{6a,5}$ 1.5	H-6a
	2.37	1	ddd	$J_{6b,6a}$ 12.5, $J_{6b,7}$ 11.0, $J_{6b,5}$ 9.0	H-6b
	2.68	1	m	$J_{5,6b}$ 9.0, $J_{5,1}$ 8.0, $J_{5,6a}$ 1.5, $J_{5,4}$ 1.5	H-5
	3.47	1	ddd	$J_{7,6b}$ 11.0, $J_{7,6a}$ 6.5, $J_{7,8}$ 4.5	H-7
	3.58	1	ddd	$J_{1,5}$ 8.0, $J_{1,3}$ 1.0, $J_{1,8}$ 1.0	H-1
	3.80	1	dd	$J_{8,7}$ 4.5, $J_{8,1}$ 1.0	H-8
	4.28	1	d	J 11.0	CHaHbPh
	4.38	1	d	J 12.0	CHaHbPh
	4.50	1	dd	$J_{4,5}$ 1.5, $J_{4,3}$ 1.0	H-4
	4.68	1	d	J 12.0	CHaHbPh
	4.77	1	d	J 11.0	CHaHbPh
	6.55	1	dd	$J_{3,1}$ 1.0, $J_{3,4}$ 1.0	H-3
	7.11-7.50	10	m		Ar-H
	9.73	1	s		CHO
m/z (+ve FAB, thioglycerol):	365 ([M + H]$^+$, 10%), 364 (M$^+$, 7), 57 (100)				
Found:	[M + H]$^+$ 365.1751. C$_{23}$H$_{24}$O$_4$ requires 365.1748				

Figure 3.5 Tabulated experimental data for inclusion in a thesis.

(±)-*exo,exo*-7,8-Dibenzyloxy-*exo*-*cis*-bicyclo[3.3.0]oct-2-en-4-ol-2-carboxaldehyde. *A stirred solution of oxalyl chloride (91 cm³, 1.0 mmol) in dry dichloromethane (15 cm³) was placed under a nitrogen atmosphere and cooled to -78 °C. A mixture of dimethylsulfoxide (182 cm³, 2.35 mmol) and dichloromethane (1 cm³) was then added dropwise. After 5 min exo,exo-7,8-dibenzyloxy-exo-3,4-epoxy-exo-2-(hydroxymethyl)-cis-bicyclo[3.3.0]octane (200 mg, 0.55 mmol) in dichloromethane (2 cm³) was added dropwise and after a further 20 min triethylamine (1 cm³, 7.61 mmol) was added. After 10 min at −78°C, the mixture was allowed to warm to room temperature, then partitioned between 2 M hydrochloric acid (30 cm³) and dichloromethane (2 × 30 cm³). The organic phase was washed with sat. aq. sodium hydrogen carbonate (40 cm³), dried (MgSO₄) and then concentrated under reduced pressure. The residue was dissolved in dichloromethane (15 cm³) and 1,5-diazabicyclo[5.4.0]undec-5-ene (167 mg, 1.1 mmol) added. The mixture was stirred at room temperature for 2 h, then poured into 2 M hydrochloric acid (20 cm³) and extracted with dichloromethane (2 × 30 cm³). The combined extracts were washed with sat. aq. sodium hydrogen carbonate (40 cm³), dried (MgSO₄) and then concentrated under reduced pressure. Purification by flash chromatography (petroleum ether/ethyl acetate, 1/1) provided the title compound (159 mg, 80%) as a colorless oil. Rf 0.3 (pet. ether/ethyl acetate, 1/1); νmax/cm⁻¹3400, 3075, 3050, 2750, 1690; δH (300 MHz, CDCl₃) 1.84 (1 H, ddd, J 12.5, 6.5, 1.5, H-6a), 2.37 (1 H, ddd, J 12.5, 11.0, 9.0, H-6b), 2.68 (1 H, m, J 9.0, 8.0, 1.5, 1.5, H-5), 3.47 (1 H, ddd, J 11.0, 6.5, 4.5, H-7), 3.58 (1 H, ddd, J 8.0, 1.0, 1.0, H-1), 3.80 (1 H, dd J 4.5, 1.0, H-8), 4.28 (1 H, d J 11.0, CHaHbPh), 4.38 (1 H, d, J 12.0, CHaHbPh), 4.50 (1 H, dd, J 1.5, 1.0, H-4), 4.68 (1 H, d, J 12.0, CHaHbPh), 4.77 (1 H, d, J 11.0, CHaHbPh),6.55 (1 H, dd, J 1.0, 1.0, H-3), 7.1-7.5 (10 H, m, Ar-H), 9.73 (1H, s, CHO); m/z (+ve FAB, thioglycerol) 365 ([M + H]⁺,10%), 364 (M⁺, 7). HRMS found [M + H]⁺ 365.1751, C₂₃H₂₄O₄ requires 365.1748.*

Figure 3.6 An example of a journal-specific experimental procedure.

Bibliography

1. IUPAC Handbook (also available online).
2. Details of data formats accepted by most major chemistry journals can be found in the "Instructions for Authors." These are usually posted on publishers' websites. Note that the styles differ slightly from one journal to another.

chapter four

Equipping the laboratory and the bench

4.1 Introduction

In this chapter, we describe general laboratory and bench equipment that we have found to be of use when performing modern organic chemical reactions. Much of the equipment introduced in this chapter will be discussed in more detail in subsequent chapters.

4.2 Setting up the laboratory

The basic furniture provided in organic chemistry laboratories will vary considerably from one establishment to another and clearly any advice given in this chapter will have to be tailored to the facilities available. The ideal layout of the laboratory is also a very subjective matter, and the advice given here is therefore not intended to be taken as gospel but simply reflects the experiences of the authors from various laboratories in which they have worked.

When setting up a laboratory, it is usual for some areas to be set aside for storage of communal apparatus and other parts to be allocated as personal bench and cupboard space for individuals working in the laboratory. Similarly, apparatus will usually be assigned as communal or individual. In this chapter, we suggest which laboratory equipment would be suitable for communal use and which would be best as part of an individual bench set, but this is only a rough guide and the arrangements in your laboratory may be different. The distinction will to some extent depend on the type of work being undertaken, and it will also depend on the budget and space available.

Most modern labs have clean write-up areas associated with them. These usually consist of separate office areas, but it is becoming increasingly common to have a glass partition between the laboratory and the associated write-up area. The latter arrangement has the advantage that it allows workers to maintain line of sight with their experiments while reading the literature and writing up data. If neither of these options is available, it is a good idea to organize some desk or bench space in the laboratory where workers can read and write, away from areas used for chemicals. It is also useful to have a communal whiteboard in this area of the laboratory to aid scientific discussion and serve as a notice board.

The area that constitutes an individual's "bench area" will vary considerably from one laboratory to another. As all procedures involving organic chemicals need to be carried out in an efficient fume hood, a full-time worker in an organic chemistry laboratory typically requires *at least* 1 m of fume hood space in order to work safely. In labs where space is at a premium, the fume hood and associated storage space may constitute the entirety of an individual's bench area. Alternatively, an individual may have additional bench-top and cupboard space within the laboratory area. In this chapter, the term bench refers to the space occupied by an individual worker and it is assumed that this space incorporates an adequate area of fume hood, as well as adequate storage space. Many modern labs now also have provision for a computer terminal or laptop PCs to be located adjacent to the fume cupboard area so that electronic laboratory notebook entries can be made at the bench. A variety of stands and switching devices are also available so that a keyboard and screen located at the laboratory bench can be linked to an office PC some distance away.

4.3   General laboratory equipment

In this section, we describe the communal facilities that will be required in a laboratory in which modern synthetic organic chemistry is performed. Careful thought should go into the placing of communal equipment. It may be quite reasonable to place unattended items such as a refrigerator or an oven in an awkward corner, but a piece of apparatus at which a person will be working should be placed in a location where the operator has enough room to work without hindering anybody else. Equipment that is used regularly by all members of the laboratory should be located in a fume hood where it is easily accessible and does not cause a disturbance to anyone working close by.

4.3.1   Rotary evaporators

Rotary evaporators are one of the most heavily used pieces of equipment in an organic research laboratory. They should therefore be located conveniently throughout the laboratory, preferably enclosed in fume cupboards or ventilated cabinets.

4.3.2   Refrigerator and/or freezer

A refrigerator and/or a freezer should be provided to store chemicals; such equipment should have no internal light (spark hazard) and should *never* be used for food storage.

4.3.3   Glass-drying ovens

Glass-drying ovens should be conveniently located around the laboratory.

4.3.4 Vacuum oven

A vacuum oven may be used infrequently, but it is still a very valuable piece of equipment, particularly for drying solid products or reagents.

4.3.5 Balances

A two-place balance is indispensable and a four-/five-place balance must be available for smaller scale work. When locating these balances, remember that they are used frequently by people carrying out careful work, so do not put them in the way of others. Also, position them so as to avoid drafts, vibrations, and magnets as these can lead to errors in weighing.

4.3.6 Kugelrohr bulb-to-bulb distillation apparatus

This apparatus is invaluable for the distillation of small quantities of high-boiling liquids. It is relatively mobile, and so it is usually stored in a cupboard and moved to a fume hood when required. When in use, the apparatus needs to be attached to a vacuum pump, so it is often convenient to locate these items close together.

4.3.7 Vacuum pumps

Various types of vacuum pumps (see also Chapter 8) are required around the laboratory. A modest vacuum source (about 20–50 mmHg) is sufficient for rotary evaporation of most solvents, filtering under suction, and distillation of relatively volatile liquids. This level of vacuum is sometimes provided by a house vacuum system; alternatively, a water aspirator or a small diaphragm pump can be used. Every rotary evaporator will need a vacuum pump of this type attached to it, and ideally there should be one of these permanently located in every fume hood.

A reliable medium vacuum (5–15 mmHg) can be provided by oil-free polytetrafluoroethylene (PTFE)-lined diaphragm pumps. These pumps are extremely useful and can often be used instead of either water aspirators or oil pumps. They circumvent the suck-back problems associated with water aspirators and have the advantage over oil pumps that they are solvent resistant and can usually be used without liquid nitrogen–trapping systems. They can be used for rotary evaporation, vacuum distillation, and many other general tasks. For most purposes, they can also be used as the vacuum source for double manifold–type inert gas lines (Section 4.4.3 and Chapter 9, Section 9.2). Most modern diaphragm pumps come with a pressure regulator that allows you to control the generated vacuum.

High vacuum oil pumps (providing a vacuum of <1 mmHg) are necessary for certain operations and these include removing the last traces of solvent from small quantities of products, degassing solvents, and

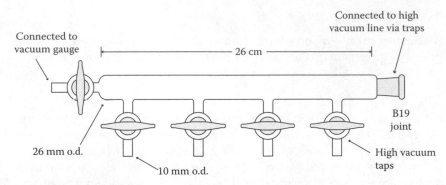

Figure 4.1 Single manifold.

distilling high-boiling compounds. They can also be used as the vacuum source for double manifold–type inert gas lines. A good two-stage rotary pump will provide a vacuum of <0.1 mmHg, which is enough for almost any task in an organic chemistry laboratory. It is very useful to have a general-purpose communal high vacuum pump permanently attached to a single manifold (Figure 4.1) that can be used for removing trace amounts of solvent from reaction products.

Ideally, a separate "high quality" pump should be reserved for distillations so that a reliable high vacuum is always obtained. These pumps should be fitted with efficient liquid nitrogen traps to avoid solvent contamination, and can be located on a mobile trolley or in a fixed position, depending on the fume hood and floor space available. A pump that is used only for distillations does not require a manifold to be attached to it.

If sufficient resources are available, each fume hood should have a double manifold connected to a medium or high vacuum pump, but in many labs this is not possible, and vacuum pumps are considered to be communal equipment.

4.3.8 Inert gases

A constant supply of an anhydrous inert gas (usually argon or nitrogen) is essential in a synthetic organic chemistry laboratory. Many labs will have nitrogen or argon lines plumbed into the fume hoods. This is ideal provided there are enough outlets and the gas is kept dry. However, in some labs gas must be supplied from cylinders, each of which must be securely fixed to a bench. It is important when setting up the laboratory to decide how many gas outlets are required; too few outlets can lead to inefficient working. We suggest that there should be at least one outlet for each person working in the laboratory plus one for each piece of apparatus (such as a still) that needs to be kept permanently under an inert atmosphere. If you are considering using inert gas cylinders in the laboratory, bear in mind that cylinder rental costs are high. One way of minimizing the number of cylinders required is to fit each one with a multiway needle valve (as discussed in Chapter 7, Section 7.2). Such valves are available from gas line hardware suppliers at a modest cost and give multiple gas outlets from each cylinder head.

Once the number of inert gas outlets required is decided, they should be positioned in the laboratory such that semipermanent polyvinyl chloride (PVC) (Tygon) tubing lines can easily be taken to each piece of apparatus requiring a supply.

4.3.9   Solvent stills

Two basic types of solvent stills are used in labs: one is for distillation of solvents for routine use and the other is for distillation of ultradry solvents for carrying out reactions under anhydrous conditions. Wherever possible you should avoid using solvent stills. This can be achieved by purchasing high purity or anhydrous solvents or by using solvent-drying towers (Chapter 5, Section 5.4.1). However, in most synthetic organic chemistry labs, there will be times when it becomes necessary to set up and use a solvent still.

If solvents are purchased in bulk (drums) or have low purity, they will normally need to be redistilled before use. Large stills (up to 5 L) may be required for this purpose. Typical solvents requiring such stills are petroleum ether, ethyl acetate, and dichloromethane. There are many problems associated with the routine distillation of all laboratory solvents. Stills are hazardous, they take up valuable space, the process is time consuming, a considerable volume of solvent is always wasted, and disposal of the residues costs money. It is generally more efficient and cost-effective to purchase bottled solvents that are sufficiently pure for routine laboratory use without distillation. We find that general purpose reagent (GPR)-grade solvents from reputable dealers are adequate for most purposes, including large- and medium-scale chromatography. Small quantities of higher purity solvents can be purchased for specific purposes and can be easily redistilled.

Some solvents such as tetrahydrofuran, toluene, and dichloromethane are frequently required in their anhydrous forms, and it is inconvenient to set up a still each time a small quantity is required for a trial reaction. The efficiency of an organic laboratory is therefore enhanced by having these dry solvents available "on-tap" either from solvent-drying towers or from permanent stills, both of which provide very effective drying and are kept under an inert atmosphere. Anhydrous solvents can also be purchased from chemical suppliers, although they tend to be very expensive and cannot generally be stored well when they have been partially used. If you have access to solvent-drying towers, they should be used wherever possible so as to avoid the hazards associated with dry solvent stills. Designs for permanent stills that can be used for drying and distillation and are compact are given in Chapter 5, Section 5.4.2.

4.3.10　General distillation equipment

Apart from standard Quickfit equipment, labs should also have some one-piece distillation kits, which allow more efficient distillation. A short-path distillation apparatus is very useful for low-hold-up, high-throughput distillations, particularly on a small scale. This apparatus is designed to minimize the area of internal glass surface that your material is in contact with, reducing both material loss and heat loss during the distillation process. One-piece distillation kits are available, but they are very expensive. They can be easily constructed by a glassblower according to the design shown in Figure 4.2. The smaller sized apparatus indicated in Figure 4.2 is useful for distilling volumes up to 25 cm^3, and the larger apparatus can be used for volumes of 25–100 cm^3. Note that distance B on the diagram is crucial as it corresponds to the length of a Quickfit thermometer from

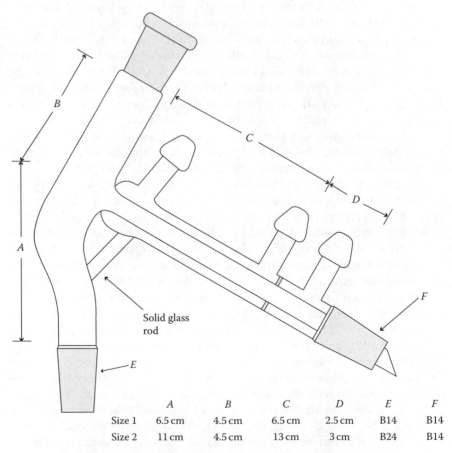

	A	B	C	D	E	F
Size 1	6.5 cm	4.5 cm	6.5 cm	2.5 cm	B14	B14
Size 2	11 cm	4.5 cm	13 cm	3 cm	B24	B14

Figure 4.2 One-piece distillation apparatus.

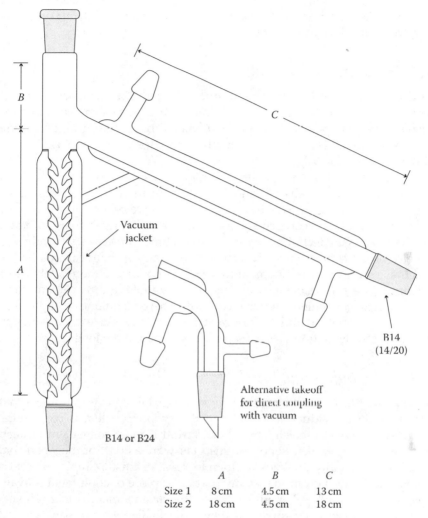

	A	*B*	*C*
Size 1	8 cm	4.5 cm	13 cm
Size 2	18 cm	4.5 cm	18 cm

Figure 4.3 One-piece distillation apparatus incorporating a fractionating column.

bulb to joint. If this is too short or too long, you need to use a thermometer adaptor that will reduce the efficiency of the setup.

For fractional distillation, a one-piece apparatus incorporating small fractionating columns can be used, and again two sizes are frequently used (Figure 4.3).

4.3.11 Large laboratory glassware

Most synthetic organic chemistry research labs routinely use glassware in the 1 cm^3 to 1 L range, but occasionally larger glassware items (e.g., round-bottom flasks, conical flasks, separating funnels, measuring cylinders) are required. As the latter will be used less frequently, it usually makes sense to have them as communal items.

4.3.12   Reaction monitoring

TLC is the most widely used technique for rapid, routine reaction monitoring, and it is essential that a laboratory has permanent facilities for visualizing TLC plates. A range of solvent dips, a heat gun, an iodine tank, and a small UV lamp will usually suffice for visualizing TLC plates (Chapter 9, Section 9.3).

Automated chromatographic techniques also are now widely used in organic chemistry labs. GC and GC-MS instruments are convenient for more volatile materials, whereas LC and LC-MS are particularly useful for monitoring routine organic chemistry experiments. UPLC-MS has become a particularly useful technique because the run times are very short and mass spectra on all components of a reaction mixture can be measured as the reaction progresses. Developments in MS techniques and the availability of relatively cheap portable instruments are now making direct MS reaction monitoring a convenient and viable option. These techniques complement TLC and, although an initial investment of time is necessary to learn how to use the instrumentation, this pays good dividends in the long run.

4.4   The individual bench

A number of factors influence the proportion of laboratory equipment that is part of an individual bench set and the proportion that is communal. In most academic research labs, workers find it most convenient to keep a more or less complete set of routine glassware as part of their personal bench kit. The individual worker then looks after this kit, replaces broken items, and keeps it clean so that any particular piece of equipment is available when required. A big advantage of this arrangement is that workers will know exactly what apparatus they have available for use and what it has previously been exposed to. In labs where all glassware is communal, there is always the risk that a particular item may not be available when you want to use it and even if the item is available it may not always be clear how thoroughly it was cleaned by the last user. In industrial labs, the situation tends to differ. There is normally an extensive range of communal routine glassware that is washed by laboratory helpers, and there is little need for workers to keep a full set of bench equipment as part of their bench kit.

In either situation, there are certain pieces of equipment that individuals usually like to keep for personal use, including things such as syringes and chromatography columns. Whether more specialized or sophisticated equipment is communal or personal will depend, to some extent, on how frequently it is used and the funds available. A list of equipment that we find useful to keep as part of the bench set is given in Tables 4.1 and 4.2, with a brief description of some of the more specialized items in Section 4.4.3.

4.4.1   Routine glassware

A typical bench set of routine personal glassware is given in Table 4.1.

Table 4.1 A Typical Set of Routine Glassware for Synthetic Organic Chemistry

Item	Quantity
Reduction adapter B14 socket B24 cone	1
Reduction adapter B10 socket B14 cone	1
Adapter tube B14 cone	1
Expansion adapter B24 socket B14 cone	2
Expansion adapter B29 socket B24 cone	2
Beaker, 10 cm^3	2
Beaker, 25 cm^3	4
Beaker, 100 cm^3	4
Beaker, 500 cm^3	2
Measuring cylinder, 10 cm^3	1
Measuring cylinder, 25 cm^3	1
Measuring cylinder, 100 cm^3	1
Measuring cylinder (stoppered), 1000 cm^3	1
Filter funnel, 7.5 cm	1
Filter funnel, 15.0 cm	1
Separating funnel, 25 cm^3	1
Separating funnel, 50 cm^3	1
Separating funnel, 250 cm^3	1
Separating funnel, 1000 cm^3	1
Conical flask, 100 cm^3	4
Conical flask, 250 cm^3	4
Conical flask, 500 cm^3	2
Round-bottom flask, 5 cm^3 B14 socket	4
Round-bottom flask, 10 cm^3 B14 socket	4
Round-bottom flask, 50 cm^3 B14 socket	4
Round-bottom flask, 100 cm^3 B24 socket	4
Round-bottom flask, 250 cm^3 B24 socket	3
Round-bottom flask, 500 cm^3 B24 socket	2
Round-bottom flask, 1000 cm^3 B24 socket	1
TLC tank	1

Table 4.2 Standard, Commercially Available Items
That Should Be Included in an Individual Bench Kit

Item	Quantity
Spatulas (various sizes)	5
Magnetic follower, 25.0 mm (standard)	4
Magnetic follower, 50.0 mm (oval)	1
Magnetic follower, 25.0 mm (oval)	4
Magnetic follower, 12.5 mm (oval)	6
Magnetic flea, 7.5 mm	6
Syringe needle, 6 in. (S-S Luer fit)	6
Syringe needle, 12 in. (S-S Luer fit)	6
Microsyringe, 100 μL	2
Microsyringe, 500 μL	2
Syringe with metal Luer lock, 1 cm^3	4
Syringe with metal Luer lock, 5 cm^3	2
Syringe with metal Luer lock, 10 cm^3	2
Syringe with metal Luer lock, 50 cm^3	1
Thermometer, −100°C to 10°C	1
Thermometer, −10°C to 110°C	1
Thermometer, −10°C to 360°C	1
NMR tube, 5 mm	5
Pasteur pipettes	1 box

4.4.2   Additional personal items

There are certain items that individuals usually prefer to keep for their
own personal use even when the routine glassware listed in Table 4.1 is
communal. A list of standard, commercially available items that we think
should be included in an individual bench kit is given in Table 4.2.

4.4.3   Specialized personal items

There are some specialized pieces of glassware that are extremely useful to have as part of the bench set, and they will be briefly described in
Sections 4.4.3.1 through 4.4.3.4, with diagrams showing how they can be
constructed by a glassblower. For full details of how they are used, see
Chapters 6, 9, 10, and 11.

4.4.3.1   Double manifold

If you wish to carry out reactions under dry and/or inert conditions, the
double manifold (Figure 4.4) is perhaps the single most useful piece of
equipment. We recommend you have this permanently installed in your

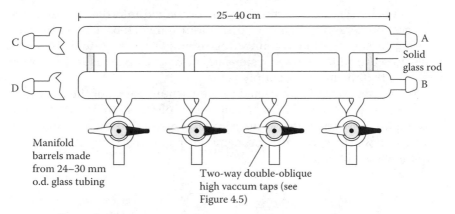

Figure 4.4 Double manifold.

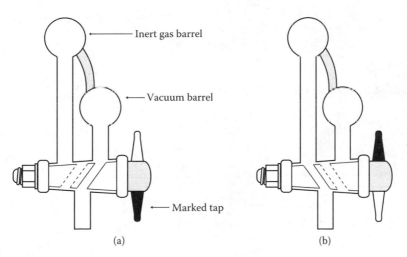

Figure 4.5 Cross section of a double-oblique tap: (a) tap switched to vacuum and (b) tap switched to inert gas.

fume hood and connected to the laboratory inert gas supply by PVC (Tygon) tubing. The manifold consists of two glass barrels, one that can be evacuated and one that is filled with inert gas. The two-way double-oblique tap (Figure 4.5) allows you to rapidly switch between vacuum and inert gas. Thus, by turning the tap, equipment connected to the manifold can be evacuated and then filled with an inert atmosphere.

A bubbler (Figure 4.6) should be incorporated into the line from the inert gas source, and it should have a built-in anti-suck-back valve to avoid the entrance of oil into the connecting tubing or the manifold. This prevents pressure buildup in the system and also provides a visual indication (bubbles) that the inert gas supply is on.

A vacuum pump is connected to the vacuum barrel of the manifold. For most purposes a PTFE-lined diaphragm pump should suffice, but if you require rigorous removal of oxygen from your reaction a high vacuum pump will be necessary. A schematic diagram showing the complete setup is shown in Figure 4.7.

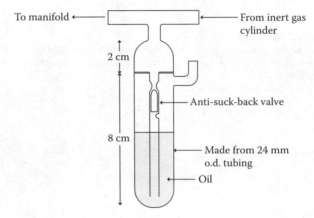

Figure 4.6 A simple bubbler design.

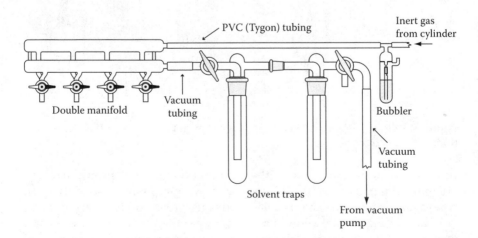

Figure 4.7 Double manifold connected to a vacuum line and an inert gas supply.

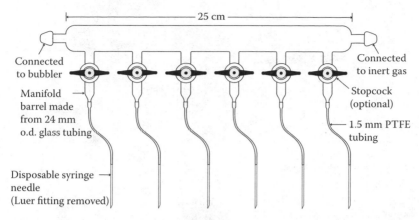

Figure 4.8 Spaghetti tubing manifold.

Although it is preferable for workers to have their own individual manifold, this may not be possible in some labs due to insufficient fume hood space or lack of vacuum pumps. In this case, manifolds must be kept as communal items of equipment. If this is the case, strict discipline must be observed to keep the manifolds clean and to avoid cross-contamination. For communal use, larger manifolds with more outlets are generally used.

Another type of manifold that is useful for carrying out reactions under an inert atmosphere is the "spaghetti tubing" manifold (Figure 4.8). This is a single-barrel inert gas manifold fitted with narrow-bore PTFE tubing outlets. Each outlet has a syringe needle attached to it that can be pushed through a septum on a reaction vessel to provide an inert atmosphere. This type of system is particularly useful for small-scale work.

For full descriptions of how to use manifolds, see Chapters 6 and 9.

4.4.3.2   Three-way Quickfit gas inlet T taps

A simple piece of equipment that is useful for a wide variety of tasks is a three-way PTFE tap connected to a Quickfit cone joint (Figure 4.9). These three-way gas inlet taps are so universally useful that we recommend you have at least three of the B14-size and two of the B24-size taps as part of your personal bench set. They are particularly useful when used in conjunction with a double manifold. With the inert gas from the manifold connected to the horizontal inlet and the tap in position A (Figure 4.10), a reaction flask can be kept under a slight positive inert gas pressure. If the gas flow is increased and the tap is turned to position B, liquids can be introduced via the vertical inlet, while maintaining an inert atmosphere.

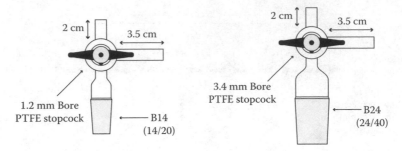

Figure 4.9 Three-way taps.

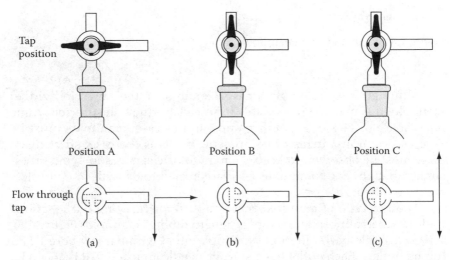

Figure 4.10 Using a three-way tap.

Another common use for three-way taps is in connecting flasks to a high vacuum system in order to remove the last traces of solvent from a sample.

4.4.3.3   Filtration aids

The rapid filtration of samples is often necessary. The speed of filtration is increased dramatically by applying pressure or a vacuum.

One-piece sintered funnels (Figure 4.11) are particularly useful for filtration under suction. The parallel-sided Buchner-style funnels can be constructed by fusing a circular sinter into glass tubing of appropriate diameter, which is then joined to a cone joint and a piece of narrow-bore tubing (about 10 mm o.d.). It is useful to have two or three of these, ranging in diameter between 1 and 10 cm, in your bench kit. The larger

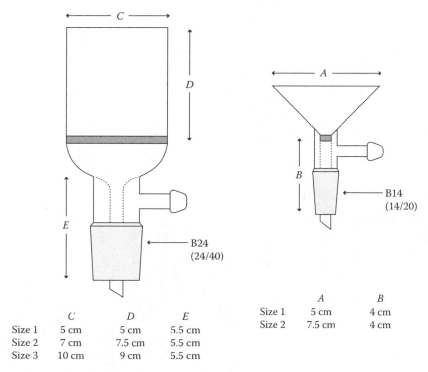

Size 1   5 cm

	A	*B*
Size 1	5 cm	4 cm
Size 2	7.5 cm	4 cm

	C	*D*	*E*
Size 1	5 cm	5 cm	5.5 cm
Size 2	7 cm	7.5 cm	5.5 cm
Size 3	10 cm	9 cm	5.5 cm

Figure 4.11 One-piece sintered filter funnels.

ones are particularly valuable for rapidly filtering off drying agents such as $MgSO_4$, allowing solutions of crude reaction products to be collected directly into a round-bottom flask attached to the B24 joint. One or two of the smaller Hirsch-style funnels are also useful, and they are more commonly used for filtering off crystals after small-scale recrystallization, the mother liquor being conveniently deposited in an attached round-bottom flask. These can be made starting from a commercial sintered funnel that lacks that B14 joint, but a glassblower can make them more cheaply by inserting a circular sinter into a narrow tube, then forming the funnel from this tube.

For small-scale recrystallization, a one-piece filtration apparatus, such as the one shown in Figure 4.12a, is very useful. It is often used in conjunction with a Craig tube (Figure 4.12b) (see Chapter 11, Section 11.2 for more details). It is likely that this type of apparatus is used only occasionally, so only one or two per laboratory should be required.

If you regularly need to filter air-sensitive solutions, or isolate air-sensitive precipitates, an inert atmosphere filtration apparatus (Figure 4.13) is essential.

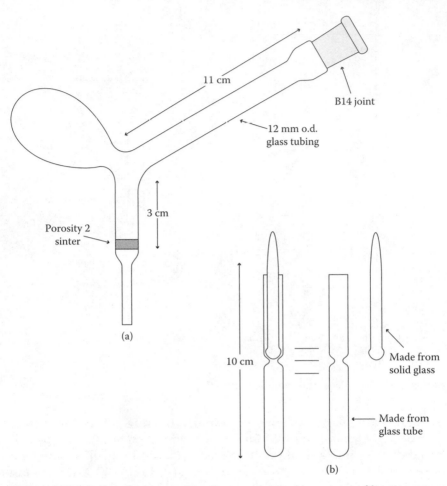

Figure 4.12 Small-scale recrystallization apparatus: (a) one-piece filtration apparatus and (b) Craig tube.

4.4.3.4 *Glassware for chromatography*

Flash chromatography is one of the most widely used methods for rapid purification and separation of reaction products. To gain expertise in flash chromatography, it is a sound idea to have a familiar set of columns (Figure 4.14c) on hand. Ideally, you should have a set of about five columns, ranging in diameter from about 5 mm to 50 mm, as part of your personal bench kit. A convenient length for most columns is 25 cm, although columns with very narrow bores may need to be shorter. Solvent reservoirs (Figure 4.14d) are also required, 250 and 500 cm^3 being the most useful sizes. Equipment with ordinary glass joints, held together with rubber bands or

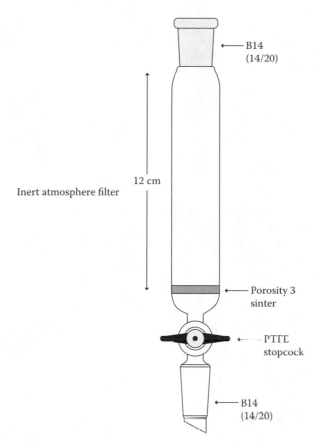

Figure 4.13 Inert atmosphere filtration apparatus.

Bibby-style clips, can be used for flash chromatography, but screw joints such as Rodaviss joints are far more reliable and safer to use. If you opt to use an inert gas source to pressurize your flash columns, a useful additional item is a simple "flash valve" (Figure 4.14b). This will regulate the pressure applied to the column, making the operation safer by reducing the possibility of pressure buildup and column fracture. A safer and more convenient means of pressurizing the column is to use either rubber bellows or a small fish-tank pump. You should be able to obtain such bellows and pumps that are able to push the solvent through the column but are incapable of generating enough pressure to risk damaging the apparatus. In this case, a simple flash adaptor (Figure 4.14a) can be used instead of the flash valve. A more detailed discussion on flash chromatography can be found in Chapter 11, Section 11.5.

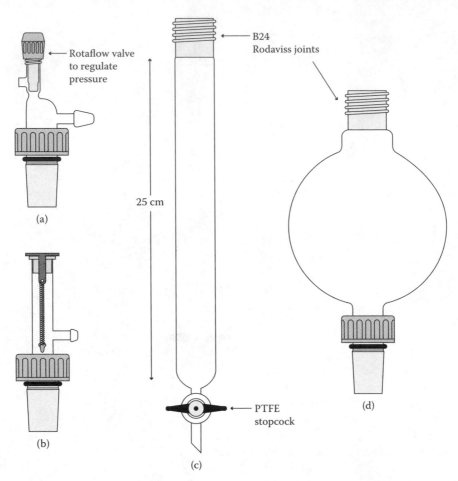

Figure 4.14 Flash chromatography column: (a) flash adapter, (b) flash valve, (c) flash column, and (d) reservoir.

4.5   Equipment for parallel experiments

In recent years, there has been a growing demand for reaction screening, particularly in the areas of catalysis and biocatalysis. The difference between an ineffective, low-yielding, and/or poorly selective reaction (e.g., low stereoselectivity and enantiomeric excess) and one that is high yielding and highly selective can be due to subtle changes in reagents used, the conditions under which the reaction is carried out, or a combination of both. Even in cases where some degree of prediction is possible, there can be a vast array of reaction parameters to be evaluated (e.g., different reagents, ligands, coreagents, solvents, concentrations, and temperatures). For this reason, many laboratories now have robotic screening equipment that enables automated or semiautomated screening of a range of reaction parameters. Robotic systems of this type can typically screen between 24 and 96 reactions in parallel. The reactions are normally carried out in multiparallel tube sets (plates), with reaction volumes of 1–20 cm^3. Some of the systems are semiautomated, where the separate tasks of weighing, reagent addition, and sampling are carried out separately. Others are fully automated "platform robots," which can be programmed to carry out a full sequence of operations automatically. As these high-throughput systems run large numbers of reactions, they also require high-throughput analysis, such as UPLC-MS. If you are working with a platform robot, you will need specific training to program and operate the equipment, and labs with this type of equipment typically have dedicated operators.

The range of equipment that is available for parallel working is wide ranging and new equipment is being introduced all the time, so it is impossible to provide a comprehensive review of such equipment in this book. However, in Sections 4.5.1 through 4.5.4, we provide a guide to some of the simpler equipment that are in common use and give an idea of the scope of their applicability. An Internet search for "parallel synthesis reactors" will provide further information on similar systems and their suppliers. Some specific equipment are referred to here, but this is not meant to be a comprehensive guide or a list of recommendations.

4.5.1   Simple reactor blocks that attach to magnetic stirrer hot plates

Magnetic stirrer hot plates (or cold plates) can be adapted so as to stir multiple flasks or tubes, and numerous simple devices have been developed to take advantage of this feature. A general limitation of these devices is that temperature control is quite rudimentary and all the reactions on a single stirrer device will be at the same temperature.

Perhaps the simplest of these devices are aluminum blocks that fit snugly on top of a magnetic stirrer hot plate and are drilled with holes that fit standard vials or small test tubes. They come in a variety of different sizes and typically take 4–24 reaction tubes or vials. The tubes or vials can be fitted with caps or septa and can be kept inert using spaghetti tubing nitrogen lines (Figure 4.8) or simply by keeping them sealed.

More sophisticated aluminum blocks are also available. For example, some have a single "cap mat" that is made of rubber and is clamped over the set of tubes, acting as a single septum for the whole set. This makes setting up and dismantling very easy. Some incorporate a water-cooled aluminum collar to cool the tops of the tubes in order to condense any vapor produced when reactions are heated. Some also have an inert gas collar that provides inert conditions with less tubing and makes assembly and disassembly easy.

Other useful variations on this type of equipment include "carousel" attachments, which are blocks engineered to take several standard round-bottom flasks, arranged in a circle. These can usually be heated or cooled and are simple to set up and very useful when you want to carry out a number of reactions under similar conditions. The reactions can be carried out in the same way as any normal round-bottom-flask reaction and can incorporate condensers, inert gas lines, and so on.

4.5.2   Stand-alone reaction tube blocks

A wide range of stand-alone thermostatically controlled heater blocks are commercially available. Such blocks typically take 6–48 reaction tubes and incorporate magnetic stirring. They are useful for reaction volumes in the range of 1–30 cm^3. Probably the best known are "STEM" blocks; they can take 10 tubes and occupy less space than a standard stirrer hot plate, which is very useful when fume hood space is limited. Commercial reaction tubes that incorporate detachable condensers are available to fit these blocks. Blocks that cool to –30°C are also available, making them useful for most types of reactions. The majority of these thermostatically controlled heater blocks will only heat or cool the entire block. This has the limitation that all the reactions in a block are performed at the same temperature. However, a number of more sophisticated reactor blocks are available today that allow the temperature of each well to be controlled independently between –30°C and 150°C. These typically take 10 reactor tubes and can be used for reaction volumes of about 1–20 cm^3. They can also incorporate an integrated computer control system that allows 10 experiments to be performed simultaneously, each with independent temperature and stirring profiles. These systems can also store and download data in a form that can be used later to gain an understanding of reactions. Not only are the systems capable of maintaining internal reactor temperatures but they can also measure the heat output that occurs during a reaction, which can be very useful in understanding the kinetics of the reaction and detecting any exotherms that are potentially hazardous.

4.5.3   Automated weighing systems

A barrier to carrying out multiparallel reactions can be the thought of repetitively weighing out a large number of samples for the experiments. Until recently, automated weighing systems were not particularly easy to use or reliable in terms of being able to weigh solids that are not free flowing. However, recent improvements in technology and software have provided a step change in usability, making modern systems a real asset for weighing multiple samples for parallel synthesis.

4.5.4 Automated parallel dosing and sampling systems

Automated liquid-handling systems that can be programmed to add reagents to reactions and/or take samples at a specified time can be very useful, particularly when carrying out design of experiments (DoE) studies. The sophisticated platform robots mentioned in Section 4.5 are able to do this, but there are also a number of simpler systems that work in conjunction with stand-alone reaction tube blocks. This type of equipment is capable of carrying out a large number of tedious manipulations, can be left running unattended (e.g., overnight), and can collect and quench sets of samples at set intervals. This can be very useful if you wish to determine reaction profiles that can be compared with each other.

4.6 Equipment for controlled experimentation

There may be times when you need to carry out reactions under carefully controlled conditions, and this is especially true when optimizing processes for scale-up. A range of equipment is now available that is designed to give precise control over single and multiparallel experiments.

4.6.1 Jacketed vessels

Jacketed vessels (Figure 4.15) are available in sizes from 100 cm^3 to tens of liters. They have an integral jacket that is circulated with fluid from a heater/chiller unit, and some vessels also have an additional outer

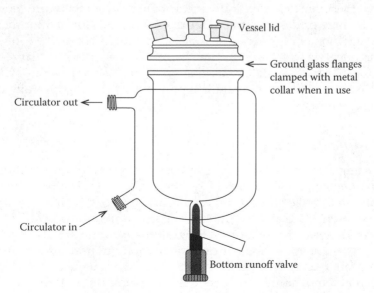

Figure 4.15 Jacketed vessel and lid.

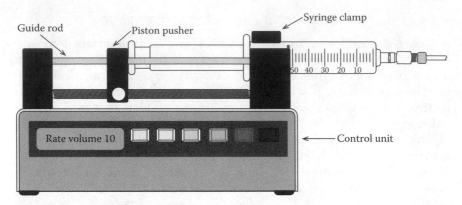

Figure 4.16 A syringe pump.

vacuum jacket. The temperature range of the setup will depend on the heater/chiller unit that is used, but typically temperatures between −30°C and 150°C are possible. Jacketed vessels normally have a separate lid, for ease of cleaning, and this will have a number of ground glass inlets. The lid is attached to the vessel by a metal collar, which clamps it in place. The most popular jacketed vessels have a bottom runoff valve, and this is very useful for removing solutions or crystallization slurries from the vessel. The bottom runoff valve also enables the vessel to be used like a separating funnel, which saves transferring reaction mixtures to a sepa-rate vessel for work-up, a particularly useful feature when working with vessels larger than 1 L. A mechanical stirrer is used for agitation, and an internal thermometer is typically incorporated for temperature monitor-ing. A syringe or dosing pump (Figure 4.16) can be attached for the con-trolled addition of liquid reagents, and other equipment can be connected via the inlets. Jacketed vessels are the reactors of choice for large-scale work (see Chapter 13) and are used at up to 100 L scale in industry. They can also be used in sets for parallel optimization studies, typically using $4 \times 250 \text{ cm}^3$ units.

4.6.2 Circulating heater-chillers

Circulating heater-chillers are very useful devices for controlling and maintaining the temperatures of reactions or crystallizations. They are available in a range of sizes and are normally able to maintain tempera-tures between −30°C and 150°C. High-capacity units are also available that can provide lower temperatures. Programmable units are also available that allow sequenced temperature changes and are useful for controlled crystallization as well as chemical reactions. Many units also come with an internal temperature probe, enabling temperature control to be based on the temperature of the circulating coolant or the internal temperature of the vessel. Circulating heater-chillers can be used in conjunction with single reactors, such as jacketed vessels, or with a range of various parallel reactor sets.

4.6.3 *Peltier heater-chillers*

Peltier heater-chiller units enable multiple reaction wells to be accurately maintained at different temperatures without the need for a heater-chiller circulator. Peltiers simply require water cooling to remove the excess heat that they generate. They have been incorporated into stand-alone reaction tube blocks and apart from maintaining individual reactor temperatures, they can also record the heat output that occurs during a reaction. This can be useful in determining the kinetics of a reaction and allows the detection of any exotherms that are potentially hazardous.

4.6.4 *Syringe pumps*

There are a range of devices available that allow precise control over the rate of addition of a reagent or substrate to a reaction. The simplest of them are syringe pumps (Figure 4.16), which are versatile devices that can be used in conjunction with any type of reaction vessel. They typically accommodate syringe sizes of 2–50 cm^3 and can be programmed to add the contents of a syringe over a prescribed period of time.

4.6.5 *Automated reaction control systems*

Most laboratory instruments (e.g., heater-chillers, syringe pumps, stirrer hot plates, and balances) are computer controlled and have the capability to be computer controlled using the output of monitoring devices, such as thermometers, pH meters, turbidity meters, and gas detectors. This means that computer-based control systems can be bought, or built, to link the operation of various devices. For example, linking a syringe pump to an internal thermometer results in a system that allows you to automatically control the rate of addition of a reagent so as to maintain a reaction mixture at a certain temperature. Similarly, linking a pH meter to a syringe pump charged with an acid or a base will allow precise control over the pH of a reaction over time. Systems of this type can be very useful if precise control over a measurable set of parameters is required.

4.6.6 *All-in-one controlled reactor and calorimeter systems*

Peltier heater-chiller control has also been incorporated into systems that are capable of being used at larger scales, and this has led to the development of sophisticated reactors that have a range of reaction-monitoring, reaction-controlling, and reagent-dosing devices. The Mettler EasyMax is an example of the type of system that is currently available. It can be used to control two 50 or 100 cm^3 reaction vessels independently or two sets of four tubes. Dedicated dosing pumps can facilitate control of reagent addition, and the reactors can be stirred either magnetically or mechanically. Peltiers enable accurate temperature control and monitoring of each reaction, and other monitoring probes, such as IR, turbidity, and pH probes, can be added. Larger scale controlled reactors are also available with reactor sizes between 250 cm^3 and 5 L, and some of them now operate with Peltier temperature control.

chapter five

Purification and drying of solvents

5.1 Introduction

The use of appropriately purified solvent(s) is vital to the success of the procedures for which they are used. It is important to note that the degree of purity and dryness required depends on the intended application, so when choosing a method from this chapter, remember to pick one that is appropriate for your purpose. Consult Appendix 1 for useful data about commonly used organic solvents.

Remember that solvents are usually hazardous materials and beware particularly of the flammability of the hydrocarbons and ethers, the toxicity of benzene, chloroform, and carbon tetrachloride, and the possibility of peroxide contamination of ethereal solvents.

5.2 Purification of solvents

The most commonly used grade of solvent is "reagent grade," which typically means 97–99% purity with small amounts of water and other volatile impurities as listed in the specification. This is adequate for use in extractions and in many chemical reactions. However, some applications are more demanding and require either the use of commercial solvents of higher purity or further purification of reagent grade products. Notable examples include the following:

1. Reactions involving strongly basic organometallic compounds, for example, Grignard reagents, organolithiums, and metal hydrides, require the use of carefully dried solvents. Anhydrous solvents (<50 ppm of water) are commercially available, but they are generally very expensive. Once opened, such commercially available dried solvents are difficult to keep dry, and so in general it is better to use solvents dried immediately prior to use using the techniques described in this chapter.
2. Reactions involving radicals, polyenes, or easily oxidized metal centers such as Pd(0) generally require solvents that have been degassed to remove dissolved O_2. This is best achieved using the

freeze-pump-thaw method. The solvent is first placed in a round-bottom flask attached to a double manifold (see Section 9.2 for details of how to do this). The flask is then placed under a positive pressure of inert gas (usually nitrogen or argon) and the solvent is frozen by immersing the flask in an appropriate cooling bath (Section 9.4). While the solvent is completely frozen, the flask is evacuated for 2–3 minutes using a high vacuum pump. The tap to the vacuum pump is then closed and the solvent is then allowed to melt while still under partial vacuum. This freeze-pump-thaw cycle needs to be repeated at least three times to ensure that all the dissolved O_2 has been removed. After the final evacuation, the flask is filled with inert gas before thawing.

3. Solvents used for spectroscopy, especially NMR and UV, should be of high purity. Many suppliers provide "spectroscopic grade" solvents that are particularly suitable for UV spectroscopy because UV absorbing impurities have been removed.

4. Solvents used for chromatography should always be fractionally distilled to ensure that nonvolatile impurities are removed. Solvents for HPLC should be of high purity and free from particulates that may cause blockages in the HPLC system. Many suppliers provide special "HPLC grade" solvents, which have been purified and filtered for this purpose. It is also good practice to degas solvents that are to be used for HPLC to prevent air bubbles from forming in the pump heads.

5. All applications involving quantitative analysis require the use of "analytical grade" (typically >99.5% purity) solvents. It is good practice in general to use high-quality solvents for the purification of your products, and it is particularly important to use pure solvents when purifying samples for microanalysis.

In most cases, purification of a solvent simply involves drying and distilling it, so Sections 5.3 and 5.4 are devoted to drying agents and methods used for drying common solvents, and Section 5.4.2 contains descriptions of typical continuous stills. Apart from water, the only other commonly encountered contaminants are peroxides formed by aerial oxidation of ethereal solvents. Methods for dealing with these dangerous impurities are described in the section on the purification of diethyl ether (Section 5.4.3).

5.3   Drying agents

Drying agents fall into two broad categories, those used for preliminary drying and the drying of extracts, and those used for rigorous drying. The pre-drying agents are largely interchangeable with one another and the choice is usually limited only by the chemical reactivity of some of the reagents. Preliminary drying of solvents, before rigorous drying, is essential unless the solvent already has low (<0.1%) water content. Of necessity, the reagents used for thorough drying are very reactive and they must be treated with great care. It is essential to make the right choice of drying agent for the solvent in question to avoid dangerous or undesirable reactions between the solvent and the drying agent, and great care must be taken to ensure the safe destruction and disposal of excess reagent remaining in solvent residues.

When the solvent is to be distilled after standing over a desiccant, the drying agent should be filtered off before distillation if it removes water reversibly, for example, by hydrate formation ($MgSO_4$, $CaCl_2$) or by absorption (molecular sieves). The solvent can be distilled without removal of the desiccant in cases where water removal is irreversible (CaH_2, P_2O_5).

The recommendations in this and Section 5.4.3 are largely based on the work of Burfield and Smithers et al., who have carried out quantitative studies on the efficiency of drying agents for a wide range of solvents.[1–7] Their work has supplanted earlier studies, many of which are of doubtful reliability. Other useful sources of information include Perrin and Armarego[8] and Riddick, Bunger, and Sakano.[9]

5.3.1   Alumina, Al_2O_3

Neutral or basic alumina of activity grade I is an efficient drying agent for hydrocarbons, 5% w/v loading giving extremely dry solvents. It is also useful for the purification of chloroform and removal of peroxides. Activated alumina is also commonly used in solvent drying towers for a range of ethers.[10]

5.3.2   Barium oxide, BaO

The commercially available anhydrous product is an inexpensive drying agent that is useful for amines and pyridines (30–50 ppm water content after standing for 24 hours over 5% w/v). It is strongly basic and is ineffective for alcohols and dipolar aprotic solvents.

5.3.3   Boric anhydride, B_2O_3

Recommended drying agent for acetone and effective also for thorough drying of acetonitrile.

5.3.4   Calcium chloride, $CaCl_2$

Both the powder and pellet forms are effective for pre-drying hydrocarbons and ethers. It reacts with acids, alcohols, amines, and some carbonyl compounds.

5.3.5   Calcium hydride, CaH$_2$

This is the reagent of choice for rigorous drying amines, pyridines, and HMPA and is also effective for hydrocarbons, alcohols, ethers, and dimethylformamide (DMF). It is available in powdered or granular form; the granular form is preferable as it can be stored for longer periods of time. The granules should be crushed immediately before use and residues should be destroyed by *careful* addition of water (H$_2$ evolution).

5.3.6   Calcium sulfate, CaSO$_4$

Also available as "Drierite," it is only suitable for drying organic extracts. The blue self-indicating version should not be used to dry liquids because the colored compound may leach into the solvent.

5.3.7   Lithium aluminum hydride, LiAlH$_4$

Although often recommended for drying ethers, it is less effective than other methods and is extremely hazardous. *Its use is strongly discouraged.*

5.3.8   Magnesium, Mg

Magnesium is the drying agent of choice for methanol and ethanol.

5.3.9   Magnesium sulfate, MgSO$_4$

The monohydrate is a fast acting drying agent and has a high capacity (it forms a heptahydrate), making it the desiccant of choice for organic extracts. It is slightly acidic so care is required with very sensitive compounds. It is usually not efficient enough to be useful for pre-drying of anhydrous solvents.

5.3.10   Molecular sieves

These are sodium and calcium aluminosilicates that have cage-like crystal lattice structures containing pores of various sizes, depending on their constitution. They can absorb small molecules, such as water, that can fit into the pores. The most commonly used types 3A, 4A, and 5A have pore sizes of approximately 3 Å, 4 Å, and 5 Å respectively, and they are available in bead or powder form. After activation at 250°C–320°C for a minimum of 3 hours, they are one of the most powerful desiccants available.[2] They can be stored in a desiccator or in an oven at >100°C for a few weeks but are rapidly hydrated in the air. If you are doubtful about the effectiveness of an old batch, place a few beads on the back of your hand and add a drop of water—if the sieves are active, they will instantly heat up as the absorption of water is exothermic. In most cases, extremely dry solvent can be obtained simply by batchwise drying over sieves, that is, allowing the solvent to stand over 5% w/v of sieves for 12 hours, decanting, adding a second batch of sieves, and so on. Sieves absorb water reversibly, so a solvent should always be decanted from the sieves before distillation. 4A beads are recommended for thorough drying of amines, DMF, DMSO, and HMPA, and almost all rigorously dried solvents are best stored over 5% w/v of 4A sieves. However, only the 3A form is suitable for drying acetonitrile, methanol, and ethanol, and higher alcohols require the use of powdered 3A sieves. They are not useful for drying acetone because they cause self-condensation. Provided that they are not discolored, sieves can be recycled by washing thoroughly with a volatile organic solvent, drying at 100°C for several hours, and then reactivating at 250°C–320°C.

5.3.11   Phosphorus pentoxide, P_2O_5

(*Causes burns*) This is a rapid and efficient desiccant, but its use is limited by its high chemical reactivity. It will react with alcohols, amines, acids, and carbonyl compounds and causes significant decomposition of HMPA, DMSO, and acetone. It is useful for drying acetonitrile and may be used for hydrocarbons and ethers but is less convenient than other reagents. It is often used in desiccators. It is best decomposed by careful portionwise addition to ice water followed by neutralization with base (do not add water directly to P_2O_5 in a glass container, the mixture may become so hot that the vessel could crack). It is extremely effective for drying gases and is available in a convenient form, mixed with an inert support, so that it does not become syrupy.

5.3.12   Potassium hydroxide, KOH

(*Causes burns*) Freshly powdered KOH is a good drying agent for amines and pyridines but is inferior to calcium hydride. It should not be used with base-sensitive solvents.

5.3.13   Sodium, Na

Sodium is widely used to dry hydrocarbons and ethers. It may be formed into wire using a sodium press or used as granules by cutting up sodium bars under petroleum ether. It suffers from the disadvantage that the metal surface rapidly becomes coated with an inert material, so it should not be used unless the solvent is pre-dried. Active sodium reacts with benzophenone to give a dark blue ketyl radical that is protonated by water to give colorless products. Thus, the sodium–benzophenone system is particularly convenient because it is self-indicating, and it is the preferred reagent in stills that are used for rigorous drying of diethyl ether, THF, 1,2-dimethoxyethane (DME), and other ethereal solvents. Sodium–potassium alloy has been recommended because it is liquid and therefore its surface does not become coated so easily, but this advantage is outweighed by the increased hazard resulting from the use of potassium. Sodium residues can be destroyed using *iso*-propanol. It is preferable to add the sodium very slowly to a large volume of the alcohol, but if this is not possible, add the *iso*-propanol slowly and carefully to the residues until hydrogen evolution ceases. The mixture should then be stirred well, to ensure that no coated lumps of sodium remain, before carefully adding methanol. After leaving for several hours, the mixture should be stirred again to ensure that all of the sodium has been consumed, and then, the mixture should be added cautiously to a large amount of water before disposal. *Sodium should never be added to chlorinated solvents because a vigorous or explosive reaction could occur.*

5.3.14   *Sodium sulfate, Na_2SO_4*

Anhydrous sodium sulfate is a weak drying agent suitable only for drying extracts. It is preferable to magnesium sulfate for drying very acid-sensitive compounds.

5.4   *Drying of solvents*

The safest way of obtaining dry solvents is to use solvent drying towers (Section 5.4.1), so where possible you should go for this option. However, if you only need a small quantity of an anhydrous solvent, or you have established that the solvent from a drying tower is not suitable, then solvents may be dried in individual batches using conventional distillation apparatus (Section 11.3) or in continuous stills (Section 5.4.2). Rigorously dried solvents must be kept under an inert atmosphere (nitrogen or argon) and handled using syringe or cannula techniques (see Chapter 6).

5.4.1   *Solvent drying towers*

Solvent drying towers are the safest way of generating anhydrous solvents and most chemistry departments and industrial companies should have a range of these available covering most of the commonly used solvents. These are usually purchased as self-contained commercial apparatus, and a typical setup for a solvent drying tower system is shown in Figure 5.1. This comprises a series of steel columns packed with an appropriate drying agent (usually activated alumina), a source of pre-dried

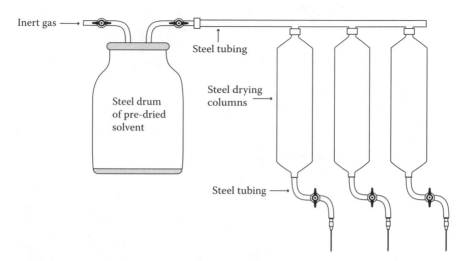

Inert gas ⟶

Steel tubing

Steel drum of pre-dried solvent

Steel drying columns ⟶

Steel tubing ⟶

Figure 5.1 Solvent drying towers.

degassed solvent, and an inert gas supply that is used to pressurize the system (typically to 15 psi) and to maintain an inert atmosphere throughout. Anhydrous solvent is then drawn off as required.

The drying columns only have a finite water capacity and so need to be regenerated on a regular basis. For volatile solvent drying columns, this is usually done by disconnecting the solvent supply, driving out all residual solvent with a continuous stream of inert gas, and then placing a heating jacket around the column and heating to >100°C with a continuous stream of inert gas until no more water vapor is expelled. Typically, this takes around 24 hours. It is essential that all the solvent has been removed before heating to avoid decomposition on the column and potential fire or explosion hazards. Nonvolatile solvent drying columns and columns that cannot be regenerated require the drying agent in the column to be replaced.

5.4.2   Solvent stills

There are two main types of solvent still that are employed in organic research laboratories. One is the classical distillation setup consisting of a distillation pot, still head, thermometer, condenser, receiver-adaptor, and collection vessel. This arrangement is described in more detail in Section 11.3. It is used for the distillation of solvents that are either required infrequently or that can be stored for long periods of time without deterioration. The other is a continuous still setup that consists of a distillation pot, collecting head, and condenser (Figure 5.2).

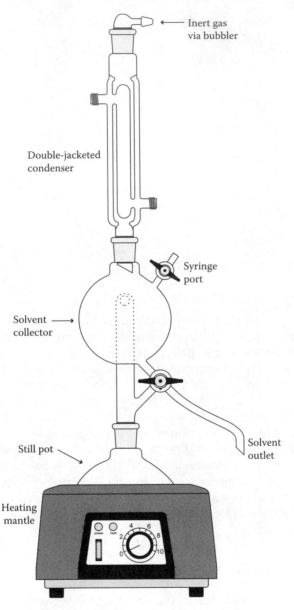

Figure 5.2 A continuous solvent still.

This type of still arrangement is used for solvents that are required on a regular basis and the still system is usually left set up, although generally it is only turned on when the solvent is required—*it is not recommended that any type of solvent still is left unattended for prolonged periods of time.*

Continuous still systems typically have an upright arrangement that takes up less space than the conventional distillation setup and a collecting vessel that is positioned between the still pot and the condenser. The apparatus is designed such that the distilling solvent is condensed and collects in a collecting head. Once the collecting head is full, the solvent simply overflows back into the still pot, allowing continuous distillation without the still boiling dry. The solvent can be drawn off from the collecting head when required or can be poured back into the still pot.

A typical design for the continuous still collecting head is outlined in Figure 5.3 and can be simply constructed from a round-bottom flask, ground glass cone, two-way tap, and three-way tap. The two-way tap allows the solvent to be withdrawn via syringe, which is particularly convenient for anhydrous solvents. The three-way tap allows the solvent to be collected, drawn off, or poured back into the distillation pot. Obviously, the size of the still depends on the quantity of solvent required; however, the still head should always be less than half the capacity of the still pot, so as to avoid the possibility of the still boiling dry.

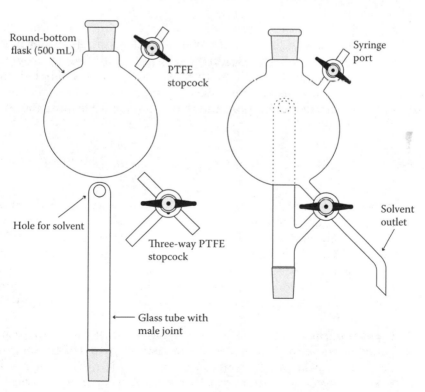

Figure 5.3 Design of how to construct a continous solvent still collecting head.

When setting up a continuous still, it is necessary to ensure that the solvent condenser is efficient; usually a double-walled condenser (Figure 9.26) is required. This is especially true for the lower boiling solvents. Also, all ground glass joints should be fitted with PTFE sleeves to ensure a good seal and prevent jamming. It is not advisable to use grease on the joints since this will be leached out by the hot solvent, contaminating the solvent and causing the joints to stick. Similarly, PTFE taps are to be preferred over glass taps in the collecting head.

If an inert atmosphere is required for the solvent distillation, as is often the case for anhydrous solvents, then the continuous still system will need to be connected to a nitrogen or argon line. It is important when using such lines that ensure that oil bubblers or similar devices do not suck back when the solvent still is cooling down. This often happens because the gas volume of the still contracts as it cools, creating a partial vacuum in the system. It is easily counteracted by turning up the flow rate of the inert gas for the period while the still is cooling down. It is also important that you do not heat up the still system without some mechanism for allowing the increase in gas volume to be released, otherwise an explosion may result. This is prevented by incorporating an oil bubbler in the gas line (Section 9.2.3).

The design of collecting heads can vary, for instance, they can also be constructed using a conical flask instead of the round-bottom flask (Figure 5.4a). Also, a more complex arrangement (Figure 5.4b) incorporating a condenser to cool the distillate (particularly useful for high boiling solvents) is sometimes used. This system also has the advantage of less ground glass joints in the setup, although this can make cleaning the still head more difficult.

5.4.3   *Procedures for purifying and drying common solvents*

1. *Acetone*

   Acetone is completely miscible with water and its susceptibility to acid and base catalyzed self-condensation makes it particularly difficult to dry. Good results are obtained by drying over 3A sieves (10% w/v) overnight (any longer causes significant condensation), stirring over boric anhydride (5% w/v) for 24 hours, and then distilling.[3] Distillation after 24 hours over boric anhydride or 6 hours over 3A sieves provides material that is adequate for most purposes.

2. *Acetic acid*

   (*Causes burns*) Acetic acid is very hygroscopic. It can be dried by adding acetic anhydride (3% w/v) and distilling (b.p. 118°C). Reagent grade acetic acid usually contains some acetaldehyde. If this needs to be removed, add chromium trioxide (2% w/v) as well as acetic anhydride before distilling or use analytical grade material.

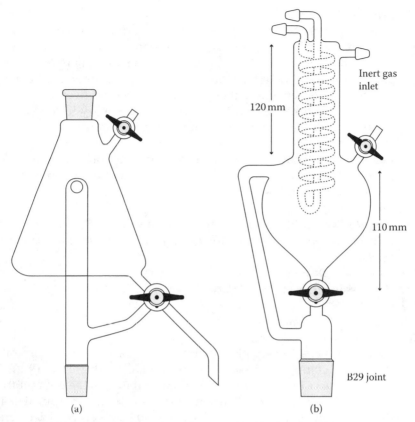

Figure 5.4 Alternative designs for solvent still collecting heads.

3. *Acetonitrile*

    (*Toxic*) Preliminary drying is accomplished by stirring over potassium carbonate for 24 hours. A further 24 hours over 3A sieves or boric anhydride gives moderately dry solvent (~50 ppm), but much better results are obtained by stirring over phosphorus pentoxide (5% w/v) for 24 hours and then distilling.[1] Drawbacks of this method are the formation of substantial quantities of colored residue and the possibility that the product may be contaminated with traces of acidic impurities. If the acetonitrile is required for use with very acid-sensitive compounds, it is best to redistill it from potassium carbonate.

4. *Ammonia*

    Distill from the cylinder into a flask fitted with a dry ice condenser (Figure 9.26) and cooled to <–40°C. Add pieces of sodium until the dark blue color persists and then distill the ammonia directly into your reaction vessel.

5. *Benzene*

   (*Carcinogenic*) Benzene, like most hydrocarbons, is very easy to dry. No preliminary drying is required and several reagents will reduce the water content to <1 ppm. Alumina, calcium hydride, and 4A sieves (all 3% w/v for 6 hours) are the most convenient drying agents and the benzene is then distilled and stored over 4A sieves.[1] Alternatively, benzene may be dried over calcium hydride in a continuous still. Toluene may be dried in the same way.

6. *tert-Butanol*

   Reflux over calcium hydride (5% w/v) and distill onto powdered 3A sieves.[7] Other low molecular weight alcohols, but not methanol, can also be dried in this way.

7. *Carbon disulfide*

   (*Highly flammable, toxic*) Distill (using a water bath) from calcium chloride or phosphorus pentoxide (2% w/v). *Do not use sodium or potassium.*

8. *Carbon tetrachloride*

   (*Carcinogenic*) See *Chloroform*.

9. *Chlorobenzene*

   See *Dichloromethane*

10. *Chloroform*

    (*Toxic*) Perhaps the simplest procedure is to pass the chloroform through a column of basic alumina (grade I, 10 g/14 cm^3). This removes traces of water and acid and also removes the ethanol that is present as a stabilizer. Carbon tetrachloride may be purified in the same way. Larger volumes of either solvent can be dried with 4A sieves or by distillation from phosphorus pentoxide (3% w/v). Distilled chloroform has no stabilizer and so should be stored in the dark to prevent formation of phosgene.

    Addition of sodium to chloroform or carbon tetrachloride may cause an explosion. Chloroform may also react explosively with strong bases and with acetone.

11. *Cyclohexane*

    See *Petroleum ether.*

12. *Deuteriochloroform (chloroform-d)*

    This is the most commonly used solvent for measurement of NMR spectra. It often contains traces of acid that can be removed by filtration through a small column of basic alumina (as for chloroform). A 6-cm plug of basic alumina in a Pasteur pipette should be sufficient to dry 0.5 cm^3 of deuteriochloroform, the amount required for a typical NMR sample.

13. *Decalin (decahydronaphthalene)*

    Decalin is very easy to dry, but it forms peroxides on prolonged contact with air, so it is advisable to use a drying agent that will destroy peroxides. Reflux over sodium for 2 hours and distill onto 4A sieves. Tetralin should be treated similarly.

14. *1,2-Dichloroethane*

    See *Dichloromethane.*

15. *Dichloromethane*

    Reflux over calcium hydride (5% w/v) and distill onto 4A molecular sieves. Chlorobenzene and 1,2-dichloroethane can be dried in the same way. Dichloromethane can be dried over calcium hydride in a continuous still.

    Never add sodium or strong bases to chlorinated solvents—an explosion may occur. Also do not mix azide salts with dichloromethane as this results in the formation of explosive azides.

16. *Diethyl ether (ether)*

    (Flammable) Ether and the other commonly used ethereal solvents (THF, DME, and dioxan) can contain substantial amounts of peroxide formed by exposure to air. *These peroxides can cause serious explosions.* Test for the presence of peroxides by adding 1 cm^3 of the solvent to 1 cm^3 of a 10% solution of sodium iodide in acetic acid. A yellow color indicates the presence of low concentrations of peroxides and a brown color indicates high concentrations. Even low concentrations of peroxides must be removed before further purification and a number of methods have been suggested.[11] Frequently recommended procedures include shaking the solvent with concentrated aqueous ferrous sulfate or passage through a column of activated alumina (which also achieves a degree of drying). Ethers are usually pre-dried over calcium chloride or sodium wire and rigorously dried over sodium-benzophenone. Careful preliminary drying is necessary because all of these solvents can dissolve substantial quantities of water. The pre-dried solvent is then placed in a reflux apparatus or a continuous still and sodium pieces (1% w/v) and benzophenone (0.2% w/v) are added. The mixture is then heated at reflux under an inert atmosphere until the deep blue color of the benzophenone ketyl radical anion persists. The ether may then be collected in a continuous still or distilled onto 4A sieves. This drying method also removes the peroxides. If a continuous still is used, it will be necessary to add more sodium and benzophenone occasionally. Eventually, the still will become very murky as the benzophenone reduction product accumulates. If this happens, or if the blue color no longer persists, it is time to clean the solvent still. To do this, distill out most of the solvent—*do not distill to dryness.* The sodium residues may then be destroyed as described in Section 5.3.13. Purified ethers are very susceptible to peroxide formation, so they should be stored in dark bottles under an inert atmosphere, and they should not be kept for more than a few weeks.

17. *DME*

    See diethyl ether. This solvent is difficult to dry rigorously using sodium-benzophenone and it is recommended that this solvent be distilled *twice* from sodium-benzophenone before use.

18. *DMF*

    Stir over calcium hydride or phosphorus pentoxide (5% w/v, overnight), filter, and distill (56°C at 20 mmHg) onto 3A sieves.[3] If phosphorus pentoxide is used, a second distillation from potassium carbonate may be necessary to remove any traces of acid. Alternatively, sequentially dry over three batches of 3A sieves (5% w/v, 12 hours).

19. *Dimethyl sulfoxide (DMSO)*

    Distill (75°C at 12 mmHg) discarding the first 20% and sequentially dry with two batches of 4A sieves (5% w/v, 12 hours).[3] Store over 4A sieves.

20. *Dioxan*

    See *Diethyl ether.*

21. *Ethanol*

    Distill ethanol from magnesium (as described for methanol) onto powdered 3A molecular sieves or sequentially dry over two batches of powdered 3A sieves (5% w/v, 12 hours).[7]

22. *Ether*

    See *Diethyl ether.*

23. *Ethyl acetate*

    Distill from potassium carbonate onto 4A sieves.

24. *HMPA*

    (Carcinogenic) HMPA is very difficult to dry and even when stored over 4A sieves it needs to be dried afresh after 2 weeks storage. Dry HMPA can be obtained by distilling (89°C at 3 mmHg) from calcium hydride (10% w/v) onto 4A molecular sieve (20% w/v).[3]

25. *Hexane*

    See *Petroleum ether.*

26. *Methanol*

    (Toxic) To dry 1 L of methanol, place magnesium turnings (5 g) and iodine (0.5 g) in a 2-L round-bottom flask fitted with a reflux condenser and then add some of the methanol (50 cm^3). Warm the mixture until the iodine disappears, and if a stream of bubbles (hydrogen) is not observed add more iodine (0.5 g). Heat the resulting mixture until all the magnesium has been consumed and then add the rest of the methanol (950 cm^3). Heat the mixture at reflux for 3 hours, then distill onto 3A sieves (10% w/v), and allow to stand for at least 24 hours before use.[7]

27. *Nitromethane*

    Pre-dry by standing over calcium chloride and then filter and distill onto 4A molecular sieves. Do not use phosphorus pentoxide.

28. *Pentane*
    See *Petroleum ether*.
29. *Petroleum ether (petrol)*
    (**Flammable**) The name "petroleum ether" is used for mixtures of aliphatic hydrocarbons containing smaller amounts of aromatic compounds. It is generally supplied as several fractions, each having a 20°C boiling range (40°C–60°C, 60°C–80°C, etc.). Alkanes that do not contain aromatic compounds are supplied as pentane, hexane, cyclohexane, and so on. All of these solvents are readily dried by distilling and standing over activity grade I alumina (5% w/v) or over 4A molecular sieves.
30. *Pyridines*
    (**Toxic**) Distill from calcium hydride onto 4A molecular sieves.[5]
31. *THF*
    See *Diethyl ether*.
32. *Tetralin (tetrahydronaphthalene)*
    See *Decalin*.
33. *Toluene*
    See *Benzene*.
34. *Xylene*
    See *Benzene*.

5.4.4   Karl Fisher analysis of water content

Karl Fischer coulometry is normally used to determine the water content in "anhydrous" solvents and is suitable for analysis of samples with water content between 10 µg and 10 mg. This is an electrochemical titration method that is based on the oxidation of an SO_2-base complex with I_2, a process that consumes one equivalent of water. It is normally performed using commercial Karl Fischer titration apparatus that measures the amount of electricity consumed to determine the water content of a sample. If you plan to use solvent drying towers or to store "anhydrous" solvents for any period of time, we recommend that you use this apparatus to check the water content of your solvents on a regular basis.

References

1. D.R. Burfield, K.H. Lee, and R.H. Smithers, *J. Org. Chem.*, 1977, **42**, 3060.
2. D.R. Burfield, G.H. Gan, and R.H. Smithers, *J. Appl. Chem. Biotechnol.*, 1978, **28**, 23.
3. D.R. Burfield and R.H. Smithers, *J. Org. Chem.*, 1978, **43**, 3966.
4. D.R. Burfield and R.H. Smithers, *J. Chem. Technol. Biotechnol.*, 1980, **30**, 491.
5. D.R. Burfield, R.H. Smithers, and A.S.C. Tan, *J. Org. Chem.*, 1981, **46**, 629.
6. D.R. Burfield and R.H. Smithers, *J. Chem. Educ.*, 1982, **59**, 703.

7. D.R. Burfield and R.H. Smithers, *J. Org. Chem.*, 1983, **48**, 2420.
8. D.D. Perrin and W.L.F. Armarego, *Purification of Laboratory Chemicals*, 3rd ed., Pergamon, Oxford, 1988.
9. J.A. Riddick, W.B. Bunger, and T.K. Sakano, *Organic Solvents, Physical Properties and Methods of Purification*, 4th ed., Wiley-Interscience, New York, 1986.
10. A.B. Pangborn, M.A. Giardello, R.H. Grubbs, R.K. Rosen, and F.J. Timmers, *Organometallics*, 1996, **15**, 1518.
11. D.R. Burfield, *J. Org. Chem.*, 1982, **47**, 3821.

chapter six

Reagents

Preparation, purification, and handling

6.1   Introduction

One of the key steps to becoming a successful practitioner of modern organic chemistry is knowing how to handle and store air- and moisture-sensitive reagents, with the certainty that they have not been contaminated. This is a skill that takes some time to acquire, and unfortunately, many people learn the hard way, after a string of failed reactions. An efficient organic chemist achieves good results rapidly, not by cutting corners but by rigidly observing strict working practices that allow sensitive reagents to be used with confidence. The biggest waste of research time stems from employing reagents or procedures that you *think* are "OK." It will usually take far more time to repeat a reaction than it would have taken to repurify a suspect reagent before starting. Even if the experiment is a success, there is always a degree of uncertainty as to whether impurities may have affected the outcome of the process. No matter how carefully the outcome of an experimental reaction is quantified, in terms of yield, stereoselectivity, by-products, and so on, the data will be meaningless if there was any uncertainty about the reaction conditions or reagents.

Handling sensitive reagents confidently is not difficult once a few standard techniques are learned and adhered to. In this chapter, we will give examples of simple general methods that can be employed to handle a wide variety of reagents.

6.2   Classification of reagents for handling

The methods used to handle a particular reagent will be dependent on the properties of that reagent and you should be fully conversant with these before you start work. If you are working with a reagent and you are not familiar with its properties, you should look them up or consult someone who is familiar with them before you begin. This is very important if you are to use the reagent effectively, without causing a hazard to you and others around you. Once you are conversant with the properties of a particular reagent, the handling requirements are largely a matter of common sense. For example, it is clearly unnecessary to rigorously exclude atmospheric moisture from a reagent that is to be used in aqueous solution, but it is important if the reagent is pyrophoric. Most reagents fall into one of the four categories as follows:

1. *Stable nontoxic reagents*
   Reagents that are neither sensitive to the atmosphere nor toxic are normally stored in ordinary bottles or containers on a shelf and they are usually straightforward to handle. However, if you are going to use this type of reagent for reactions with air-sensitive materials, *they must also be dry and be stored under an inert atmosphere*. It is pointless carefully setting up an anhydrous reaction and then adding a reagent straight from an unsealed bottle on the shelf.
2. *Stable reagents that are toxic or have an unpleasant odor*
   These should be treated in the same way as those in the preceding category, except that special precautions should be taken to avoid their escape into the laboratory. They should always be used in a fume hood and stored in a ventilated storage cupboard. When using materials of this type, it is essential that you are familiar with safe procedures for disposing of excess reagent before starting the experiment.
3. *Reagents that will decompose on exposure to moisture or air*
   These reagents should always be stored in special containers, under a dry inert atmosphere. Whenever they are used, they should be measured and transferred using techniques that constantly maintain the inert atmosphere. Some of these techniques are described in Section 6.4.
4. *Reagents that decompose explosively or pyrophorically on exposure to moisture or air*
   These should be treated as in the preceding category, but extra care should be taken, especially when dealing with residues once the reaction is over.

6.3   Techniques for obtaining pure and dry reagents

Before using any organic reagent or starting material in a reaction, its purity should be checked. It should not be assumed that the reagent is pure simply because it has been obtained from a commercial source. Indeed, the specification of many commercial compounds is much less than 100% and so this should always be checked. Even if the specified purity of the reagent is high, at least one analytical check should also be carried out (e.g., TLC, HPLC, or NMR). This check is particularly important if you are attempting a new reaction. If the purity of the reagent is not up to the level required, standard purification techniques (distillation, recrystallization, chromatography, sublimation, etc.) should be used (see Chapter 11 for more details).

Remember that any reagent that is to be used in a reaction under inert conditions with moisture-sensitive reagents must be rigorously dried before use.

Once you have taken the trouble of purifying and drying a reagent, it is good practice to store it very carefully to keep it dry for future use. This normally means storing it in a sealed container, under an inert atmosphere. When you keep reagents in this way, you must be able to rely on their purity, and it is therefore best to keep them for your own personal use. A person who does not want to take the trouble of purifying a reagent for themselves is probably not going to take particularly good care of yours.

In this section, we will describe some typical techniques for purification and drying of reagents. There are several other more specialized texts that should be consulted if you need to purify other reagents.[1-3]

6.3.1   Purification and drying of liquids

For most liquids, the method of choice for both purification and drying is distillation. In many cases, the liquid is either dried over a drying agent before distillation or distilled from a drying agent (see also Chapter 5).

1. *Distillation under inert atmosphere at normal pressure*
   This technique is used to purify and/or dry most liquid reagents that boil at less than 150°C at atmospheric pressure. Typically, the liquid is pre-dried by shaking over magnesium sulfate and then decanted onto a more active drying agent, such as powdered calcium hydride, from which it is distilled. It is important to make sure that the drying agent does not react with the reagent. If you require a dry reagent, it is useless to carry out the distillation in the atmosphere; the process must be carried out under an inert atmosphere (argon or nitrogen). For ease and reliability, we suggest that a one-piece distillation apparatus is used for this type of distillation, in conjunction with a double manifold as described in Section 11.3.

   When the distillation is complete, remove the collector, seal it quickly with a septum, and purge the bottle with inert gas. For compounds that react with rubber (e.g., Lewis acids), a PTFE stopper should be used. Some of the less sensitive reagents can then simply be poured into a reagent bottle before it is sealed, provided this is done quickly. However, if the reagent is particularly sensitive to air or moisture (e.g., Lewis acid), a cannulation technique should be used to transfer it to another container. See Section 6.4.1 for more details on how to store and transfer reagents under inert conditions. Table 6.1 gives some examples of common reagents that should be distilled under an inert atmosphere.

2. *Distillation under reduced pressure*

For liquids with a higher boiling point, or which decompose on heating to their boiling point, distillation under reduced pressure is required. Again, it is generally convenient and effective to use a one-piece distillation apparatus in conjunction with a double manifold for this type of distillation. The procedure is described in more detail in Section 11.3.4, and some examples of reagents that can be distilled under reduced pressure are given in Table 6.2.

When the distillation is complete, the heat is turned off and the two-way tap on the double manifold is slowly turned to the inert gas position. The flask containing the distillate will then be under an inert atmosphere and should be quickly removed and fitted with a tightly fitting septum.

3. *Some special cases for drying of liquids*

There are a number of liquid reagents that hydrolyze when they come in contact with water and so are frequently contaminated with the hydrolysis product, which is often an acid. These reagents are typically distilled from a high boiling point amine, or other base, so that any acidic impurity is removed. It is always important that the distillation is carried out under inert atmosphere (or under reduced pressure) using the techniques mentioned earlier. Some examples of reagents that fall into this category are given in Table 6.3.

Table 6.1 Examples of Reagents That Should Be Distilled under an Inert Atmosphere

Reagent	Drying agent	B.p. (°C)	Comment
Methyl iodide	$CaCl_2$ before dist.	43	Very toxic! Wash with $Na_2S_2O_3$ or pass through alumina first to remove I_2
Diisopropylamine	Dist. from CaH_2	84	
Triethylamine	Dist. from CaH_2	89	
Isobutyraldehyde	$CaSO_4$ before dist.	62	Readily oxidized
Titanium tetrachloride	None	136	Highly corrosive, very reactive toward moisture
Cyclohexene	Dist. from Na	83	Wash first with $NaHSO_3$ to remove peroxides
Crotonaldehyde	$CaSO_4$ before dist.	104	Use Vigreux dist. column

Table 6.2 Examples of Reagents That Can Be Distilled under Reduced Pressure

Reagent	Drying agent	B.p. (°C/mmHg)	Comment
Dimethyl formamide	$MgSO_4$ before dist.	76/39	Do not use CaH_2 or other basic drying agents
$BF_3 \cdot Et_2O$	Dist. from CaH_2	67/43	Reacts with moisture
Et_2AlCl	Dist. from NaCl	107/25	*Spontaneously flammable in air*
Et_3Al	None	130/55	As for Et_2AlCl
Benzaldehyde	$MgSO_4$ before dist.	62/10	Wash with Na_2CO_3 before dist., store over 0.1% hydroquinone
Benzyl bromide	$MgSO_4$ before dist.	85/12	Dist. in dark; *highly toxic, lachrymator*

Table 6.3 Examples of Reagents That Can Be Distilled from Quinoline

Reagent	Distil from	B.p. (°C)	Comment
Acetic anhydride	Quinoline	138	10:1 with quinoline
Acetyl chloride	Quinoline	52	10:1 with quinoline
Me_3SiCl[a]	Quinoline	106	10:1 with quinoline

[a] Another method for removing HCl from trimethylsilyl chloride (chlorotrimethylsilane) is to add it to an equal volume of triethylamine in a centrifuge tube, sealed under argon with a septum. Centrifugation then compacts the insoluble triethylamine hydrochloride at the bottom of the tube, and the clear liquid, which is approximately 50% Me_3SiCl, 50% Et_3N, can be drawn off by syringe and used as is for most purposes.

6.3.2 Purifying and drying solid reagents

If the purity of a solid reagent is not at the level required, standard purification techniques (chromatography, recrystallization, and sublimation) need to be used (see Chapter 11 for more details).

Remember, if a solid reagent is to be used in a reaction under inert conditions with moisture-sensitive reagents, it must be dried very carefully. For bulk drying of solids, an oven is often used (preferably a vacuum oven), or the solid is placed in a vacuum desiccator over an appropriate drying agent (P_2O_5, H_2SO_4, etc.). If the reagent is to be stored for future use in reactions under inert conditions, it should be kept under argon (or nitrogen) in a sealed bottle. Small quantities of solid can often be dried directly in the reaction flask, before setting up the apparatus. This is conveniently done by connecting a dry reaction flask containing the solid to a double manifold, evacuating under high vacuum (with heating where necessary) for several hours, and then introducing argon (or nitrogen).

Purification of some common solid reagents

1. *α,α'-Azobis(isobutyronitrile) (AIBN)*
   Recrystallize from diethyl ether and dry under vacuum over P_2O_5 at room temperature. Store under inert atmosphere, in the dark, at –10°C.

2. para-*Toluenesulfonyl chloride*
   para-Toluenesulfonyl chloride often contains a considerable quantity of *para*-toluenesulfonic acid. This can be removed by placing the impure reagent in the thimble of a Soxhlet apparatus (Figure 10.2) and extracting with dry petroleum ether. After several hours of extraction under an inert atmosphere, the chloride will dissolve in the petroleum ether and the unwanted acid will be left behind in the Soxhlet thimble. On cooling the solvent mixture, the acid chloride crystallizes and can be collected by filtration. The purified material should be stored under an inert atmosphere.

3. *Copper(I) iodide (for the preparation of cuprate reagents)*
   Copper(I) iodide is often slightly brown because of contamination by iodine and copper(II) salts. It is very important that these contaminants are removed and that the reagent is dried efficiently if it is to be used in the preparation of copper–lithium (cuprate) reagents. This is accomplished by placing the material in the thimble of a Soxhlet apparatus and extracting with dichloromethane for several hours (usually overnight) until no further color is being removed. The almost white solid is then collected from the thimble, dried under vacuum, and stored under argon (or nitrogen) in a dark bottle.

4. *Magnesium (for Grignard reactions)*
   Magnesium turnings should be washed with diethyl ether to remove grease from the surface of the metal and then dried at 100°C under vacuum. For a more active form of magnesium, stir the washed, dried turnings under nitrogen overnight. They will turn almost black as the oxide coat is removed from the surface. Magnesium activated in this way should be used immediately, preferably in the flask in which it is generated.

5. *Zinc*
   Zinc is normally coated with oxide, which needs to be removed before use. This can be done by stirring with 10% aqueous HCl for 2 minutes and then filtering and washing with water, followed by acetone. The metal can then be vacuum-dried and kept under an inert atmosphere before use.

6.4   Techniques for handling and measuring reagents

Bottles and flasks containing liquids or solutions sealed under an inert atmosphere are widely used in modern organic chemistry. Examples of such reagents are organolithium reagents, lithium aluminum hydride and other active hydride reducing agents, and Lewis acids. Extreme care must be taken to exclude air and moisture from these reagents. However, if you become familiar with the techniques outlined in this section, you should find them straightforward. When handling air- or moisture-sensitive reagents, *always think ahead* and design the whole sequence of events you intend to follow so that air and moisture never comes into contact with the reagent.

Procedures for preparation and titration of alkyllithium reagents are given in Section 6.5 and a method for titration of hydride reagents is given in Section 7.4.4.

6.4.1   Storing liquid reagents or solvents under an inert atmosphere

After drying and/or distillation, most reagents or solvents are best stored under argon, in a bottle or flask sealed with a septum or PTFE stopper. Commonly used reagents that are extremely reactive toward water and/ or oxygen are often stored in the same way. Examples of such reagents are alkyllithium reagents, Grignard reagents, organoboranes, metal hydrides, organoaluminum compounds, and Lewis acids. Some of these are available commercially and are delivered under inert atmosphere in sealed bottles. Less common reagents may be prepared in the laboratory and stored for future use. The seals of the commercially supplied containers have a limited life span, and if one is suspect, it should be replaced with a rubber septum. For Lewis acids, which often react with rubber, a PTFE stopper can be used. For storing small quantities of corrosive liquids under inert atmosphere, "Mininert valves" are very useful. These are screw-threaded bottle caps that incorporate PTFE valves through which a syringe needle can be inserted and provide a better seal than ordinary septa.

To prepare a storage vessel or reaction flask for introduction of an air- or moisture-sensitive reagent:

1. Dry the bottle or flask thoroughly, either in an oven or by heating under vacuum, and then cool under an inert atmosphere.
2. Select a septum or Mininert valve that fits tightly into the neck of the bottle or flask that you wish to seal.
3. Add the reagent to the bottle or flask and then fit the septum or valve as quickly as possible, minimizing exposure of the reagent to the atmosphere.

4. Insert a needle through the septum to act as a vent.
5. Quickly flush inert gas into the bottle or flask using a needle connected to an inert gas line by a Luer adapter (Figure 6.1a).
6. Once the bottle or flask has been thoroughly purged with inert gas, remove the vent and then the inert gas needle.

For long-term storage, the septum should be secured with copper wire. For extra protection, another septum of the same size, without any needle holes in it, can be turned upside down and pulled over the one that is attached to the container (Figure 6.1b and c). The second septum is removed before syringing liquid from the vessel. It is also a good idea to wrap PTFE tape, followed by Parafilm around the septum, especially if the bottle is to be stored in the refrigerator. Moisture-sensitive reagents can be added to the vessel by cannulation as described in Section 6.4.2.

A reagent or dry solvent stored under an inert atmosphere as described in Section 5.3 and used carefully according to the techniques described in Section 6.4.2 can be kept for several months and used many times over. If a reagent bottle has been stored in a refrigerator or freezer, always allow it to warm up to room temperature and make sure to remove any condensation from the outside of the container before unsealing it.

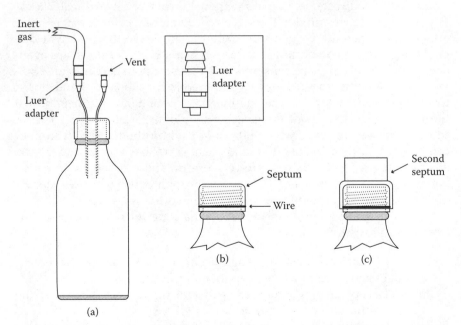

Figure 6.1 Preparing a vessel for storage of air- or moisture-sensitive reagents: (a) flush the container with inert gas, (b) wire on a septum, and (c) use an "inverted" second septum as a seal.

6.4.2 Bulk transfer of a liquid under inert atmosphere (cannulation)

As a general rule, never attempt to remove liquid from a container that is sealed under inert gas, unless you have pressurized the container with inert gas first (Figure 6.2a). Also, whenever you are using ground glass joints connected under pressure, always secure them with either plastic (Bibby-type) clips, elastic bands, or springs.

Cannulation is a general procedure for the transfer of a liquid from one container to another while maintaining an inert atmosphere throughout. The principle of cannulation is very simple. A positive pressure is applied to the container from which the liquid is to be transferred. This pressure forces the liquid out through a double-ended needle (cannula) into the receiving flask (Figure 6.3). For the liquid to flow, there must be some means by which the gas in the receiving vessel can escape. The simplest way to allow this to happen is to vent the receiving flask with a short needle passed through the septum. It is good practice to connect the vent needle to a bubbler to prevent the air being sucked back.

It is quite common that the flask into which you need to cannulate the liquid is already part of an inert gas system and cannot simply be vented. In this case, the pressure applied to the first flask must be higher than that in the receiving flask and *there must be some means by which the gas can escape from the receiving flask.* The inert gas system to which the

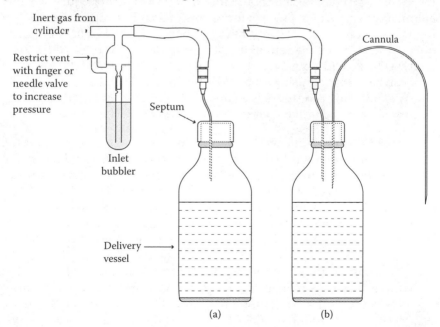

Figure 6.2 Setting up a system for bulk transfer of a liquid under inert atmosphere: (a) apply a positive pressure of inert gas and (b) flush the cannula with inert gas.

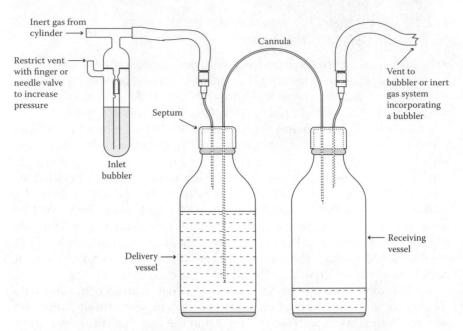

Figure 6.3 Bulk transfer of a liquid under inert atmosphere.

receiving flask is connected should incorporate a bubbler (Figure 6.3); this automatically provides a means for escape of gas. To temporarily create a higher pressure in the delivery flask or control the rate of flow of liquid during cannulation, the vent of the bubbler in the delivery system can be restricted (e.g., with your finger). For more precise control of pressure, the vent can be restricted using a Rotaflow tap or needle valve.

The following procedure is a general method for transferring liquids by cannulation and will be referred to in Chapters 9, 11, and 13.

1. Make sure the bottle or flask into which you are going to transfer the reagent is thoroughly dry, fit a septum into the neck, and purge the container with inert gas as described in Section 6.4.1. If the container was oven-dried, it is preferable to fit the septum while it is still hot and allow it to cool as it is being purged with inert gas (Figure 6.1a).
2. When the bottle or flask is cool, make sure the septum is fitted tightly, wire it on, and seal the join with PTFE tape followed by Parafilm. Then, remove the vent *before removing* the inert gas line (Figure 6.1). The container now contains an inert atmosphere and is ready for use.
3. Insert a needle connected to an inert gas line into the septum of the bottle or flask containing the liquid to be transferred. If you are transferring liquid to a flask that is already part of an inert gas system, a separate gas line must be used to pressurize the bottle and a simple bubbler system is recommended (Figure 6.2a).

4. Insert a double-ended needle (cannula) into the bottle containing the reagent, *taking care not to push the needle below the surface of the liquid at this time.* Check that the inert gas is flowing through the system and out through the cannula (Figure 6.2b).
5. Insert the other end of the cannula through the septum of the new bottle or flask (receiving flask) and then push the inlet end below the surface of the liquid in the delivery vessel. There should be no flow initially because the receiving flask is sealed. Vent it by inserting a short needle through the septum. The vent needle can be connected to a bubbler or an inert gas system that incorporates a bubbler. If the liquid does not start to flow, increase the pressure by restricting the vent of the inlet bubbler using a finger, a septum, or by connecting a needle valve (Figure 6.3).
6. When all the liquid required has been transferred, first remove the vent needle and then remove the cannula at the receiver end. The remainder of the system can then be dismantled in any order, but remember to be very careful with the residues, particularly if the liquid is toxic or reactive to moisture. Take extreme care when transferring very reactive reagents, such as tert-butyllithium, which is pyrophoric on contact with air and moisture.

6.4.3   Using cannulation techniques to transfer measured volumes of liquid under inert atmosphere

The cannulation technique can easily be adapted for measuring volumes of liquid under inert atmosphere. It is mainly used for measuring quantities of liquid that are too large to be conveniently handled by syringe (typically >50 cm^3), but it is also convenient for measuring precooled solutions without significant warming. The technique is identical to that described in Section 6.4.2, but an intermediary graduated container is used. The graduated container can be a measuring cylinder with a neck that can be fitted with a septum (Figure 6.4a) or it can be a graduated Schlenk tube (Figure 6.4b).

The apparatus is used as follows:

1. Dry the graduated container and transfer the required quantity of liquid to it according to the procedure described in Section 6.4.2 (Figure 6.4a).
2. Once the required volume of liquid has been transferred, remove the needle and bubbler from the storage bottle and insert it through the septum of a dry receiving vessel that has been dried and filled with inert gas as described in Section 6.4.1.
3. Apply inert gas pressure to the graduated cylinder and vent the receiving flask to deliver the liquid (Figure 6.5).

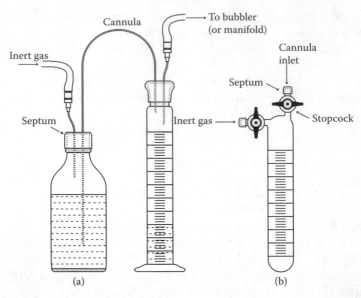

Figure 6.4 Measuring large volumes of liquid under inert atmosphere using either (a) a measuring cylinder or (b) a Schlenk tube.

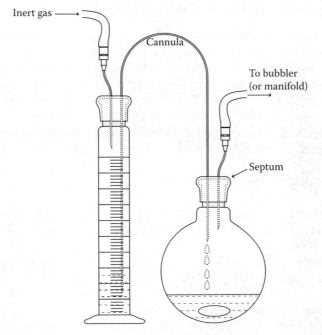

Figure 6.5 Bulk transfer of measured volumes of liquid under inert atmosphere.

If the liquid is being measured for addition to a reaction flask, an alternative procedure is to use a graduated pressure-equalizing dropping funnel attached to the apparatus. The required quantity of liquid can then be cannulated into the dropping funnel directly.

For most purposes, cannulation can be carried out using an ordinary double-ended needle, bent to a suitable "U" shape (Figure 6.6a). A cannula can also be made by joining two long syringe needles to a Luer to Luer stopcock (Figure 6.6b). This has the advantage that the stopcock can be used to control the flow of liquid. For transferring large volumes of liquid, a commercially available "flex-needle" (Figure 6.6c) can be useful. This is a jacketed PTFE tube with a wide-bore needle attached to either end. It allows rapid liquid transfer and a degree of insulation for a cold reagent. The wide-bore needles of the flex-needle should not be inserted through a new septum without first making a hole with a normal gauge needle.

For small-scale work, a PTFE cannula is very useful, and some people prefer it to syringes. They are easily made by taking a piece of narrow-bore (1.5-mm) PTFE tubing of a suitable length and cutting each end at a shallow angle (Figure 6.7) with a sharp razor blade or scalpel. This can then be inserted directly through a septum as long as you have made a hole first using a metal needle.

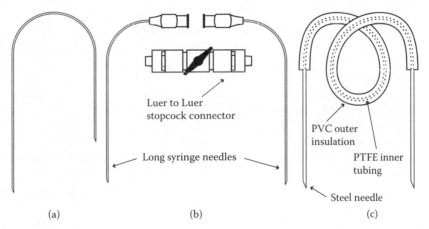

Luer to Luer
stopcock connector

PVC outer
insulation

Long syringe needles

PTFE inner
tubing

Steel needle

(a)   (b)   (c)

Figure 6.6 Different types of cannula: (a) a standard cannula, (b) a cannula made from two syringe needles and Luer locks, and (c) a "flex-needle."

Ends cut at an angle
using a scalpel

Narrow bore
PTFE tubing

Figure 6.7 Making an all-PTFE cannula.

6.4.4   *Use of syringes for the transfer of reagents or solvents*

Syringes are extremely useful for transferring small quantities (up to 50 cm^3) of air-sensitive reagents or dry solvents from one bottle or flask to another. There are a variety of syringe types and you need to appreciate which type is suitable for a particular application. It is most important to choose a syringe of an appropriate size for the quantity of liquid you need to transport. Using a syringe that is too large will lead to inaccuracy and using a syringe that is too small will waste time. For most purposes, the syringe should be equipped with a long needle (10–20 cm). You should choose a length that is sufficient to reach the reagent or solvent in a bottle without having to tilt the container. When using syringes, the condition of septa should be checked regularly. Their life can be extended by making sure that the needle tips are kept sharp by using Emery paper.

The skill of using syringes is of great importance to an organic chemist, but, as with all skills, you will only become proficient after a certain amount of practice. It is a good idea to have your own set of syringes appropriate for the type of work in which you are engaged. They will be some of your most frequently used tools, so you should practice using them and take great care of them.

There are a number of different types of syringe and fitting:

1. *Microsyringes*

   Microsyringes are extremely useful for small-scale synthetic work, delivering small quantities of liquid accurately. Sizes ranging from 10 μL to 1.0 cm^3 are common, but 100 μL is perhaps the most useful size.

   There are two common types of microsyringe: gas-tight and liquid-tight. The liquid-tight syringes have *matched* glass barrels and plungers (Figure 6.8). These are not interchangeable, so it is essential that you keep the two components together.

   These are delicate and quite expensive items, which should be treated with great care and cleaned immediately after use. Many types of reagent will corrode the steel plunger and cause seizure of the syringe if this is not done.

   Gas-tight microsyringes have PTFE-tipped plungers (Figure 6.9), making them more inert to corrosive chemical reagents (e.g., Lewis acids). The barrels and plungers are usually interchangeable and they tend to be more reliable than the liquid-tight syringes. The drawback with gas-tight syringes is that they tend to be more expensive.

   Both types of microsyringe are available with either fixed or removable needles, but those with removable needles are generally preferred. Their advantages are that the needle can be replaced when damaged or blocked and that different sized needles can be attached as required.

2. *All-glass Luer fitting syringes*

All-glass Luer syringes (Figure 6.10) typically range in size from 1 cm³ to 100 cm³ and are generally quite cheap. Some have matched a barrel and plunger and can only be used as a pair, but most modern syringes have interchangeable barrels and plungers. Syringes with interchangeable parts will be marked as such and these are the preferred types to use. Rocket-type syringes have metal plungers and are somewhat more reliable than the all-glass type. Both types are adequate for most purposes but *always* should be cleaned immediately after use and treated carefully to prevent seizing up.

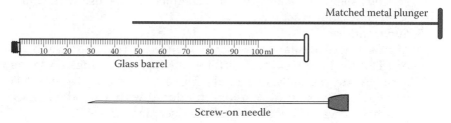

Figure 6.8 Liquid-tight syringe.

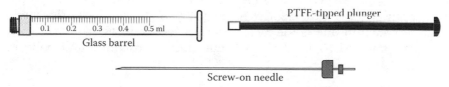

Figure 6.9 Gas-tight microsyringe.

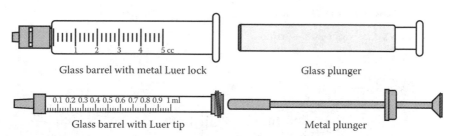

Figure 6.10 All-glass Luer syringes.

3. *Glass gas-tight Luer syringes*

For most purposes, these are the best types of syringes available. The barrel of the syringe is made from glass and the plunger is made from metal or plastic and has a separate PTFE tip (Figure 6.11). Their main advantages over all-glass syringes are that they are less prone to leakage and jamming. They are very expensive, but all the components are replaceable, so they will last a long time if they are looked after.

4. *Plastic disposable syringes*

Plastic syringes hold a good deal and are cheap, quite accurate, and virtually unbreakable. They can be used many times over but have the drawback of being susceptible to attack from some solvents and reagents. For this reason, they are most widely used for transferring aqueous solutions. They are not recommended for use with small-scale anhydrous reactions.

5. *Syringe fittings*

Microsyringes normally have screw-on needles (Figures 6.8 and 6.9) or PTFE tip Luer lock fittings (Figure 6.12). The fittings on larger syringes are normally one of the three Luer types (Figures 6.11 and 6.12). A Luer fitting is a joint produced with a standard taper, which matches the internal taper of a Luer syringe needle.

The simplest of these is the Luer glass tip, but these are easily broken and so should be avoided. Metal Luer tips are more robust, but attached needles can work loose, and for this reason, Luer lock syringes are generally preferred. These have a slot into which the rim of the needle engages so that it cannot come loose. The most common fitting is the metal tip Luer lock, but the best seal of all is provided by the PTFE tip Luer lock, normally found only on

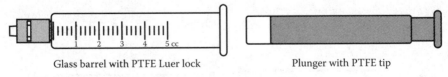

Glass barrel with PTFE Luer lock   Plunger with PTFE tip

Figure 6.11 Gas-tight Luer syringe.

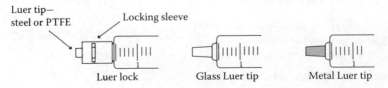

Luer tip—
steel or PTFE   Locking sleeve

Luer lock   Glass Luer tip   Metal Luer tip

Figure 6.12 Luer syringe fittings.

gas-tight syringes. In this case, the Luer tip is PTFE and the locking sleeve is metal.

6. *Syringe needles*

Luer fitting syringe needles (Figure 6.13) are available in a range of lengths and internal bore diameters. The optimum bore diameter depends on the volume of liquid being dispensed. As a rough guide, we recommend 22-gauge needles for volumes of less than 1 cm^3, 20 gauge for volumes 1–5 cm^3, and 18 gauge for larger quantities. Long needles are usually preferred for transferring liquids by syringe, but short needles are useful as nitrogen inlets or vents. Disposable needles with plastic shanks can also be used for this purpose.

Most needles have a 12° bevel and this should be kept in good, sharp condition so that minimal damage is caused when penetrating a septum. Needles can be sharpened by rubbing with fine Emery paper. Flat-ended needles are sometimes useful for removing the last traces of liquid from a container, but they should only be introduced through septa that have previously been pierced by a sharp needle.

Syringes and needles can give good service for long periods, but careful treatment is required to prevent leakage and jamming. It is therefore very important to clean syringes and needles *immediately* after use. Organic liquids or solutions are generally quite easily removed, simply by flushing with inert solvent. However, more care is required when a reactive reagent such as n-butyllithium or titanium tetrachloride has been used. To clear such reactive reagents, the syringe should *immediately* be flushed several times with an inert solvent. Whatever the syringe has been used for, it should *always* be dismantled for final cleaning of the individual components. If the syringe has been used for an alkyllithium or any other strongly basic reagent, it should be washed sequentially with dilute hydrochloric acid, water, and then acetone. After using an acidic reagent or a Lewis acid, the components should be washed sequentially with dilute aqueous sodium hydroxide, water, and then acetone.

If a syringe has seized up or become contaminated with remnants of a stubborn reagent, handle it carefully. *Do not use force* to free the plunger as this will almost certainly lead to the syringe being broken and you may

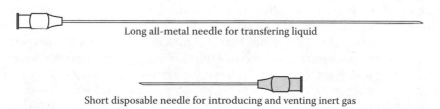

Long all-metal needle for transfering liquid

Short disposable needle for introducing and venting inert gas

Figure 6.13 Luer fitting syringe needles.

well end up with a cut hand. One of the most reliable methods for removing a contaminant from a syringe, or syringe needle, is to submerge it in a solvent that will dissolve the contaminant and to place it in an ultrasonic cleaning bath for a few minutes. Sonication will often free syringes that are completely seized up provided you choose the correct solvent. If you do not know what is blocking the syringe, try sequentially sonicating in the following series of solvents: 10% hydrochloric acid, then water, then dilute NaOH, then water, then acetone, then methanol, and finally dichloromethane.

The type of reagent that you will need to deliver by syringe is often one that has to be measured precisely for the reaction to be successful. It is also likely to be a corrosive or moisture-sensitive reagent, such as n-butyllithium, which you do not want to splash around the laboratory or spill onto your skin. So, if you have not used a syringe before, practice by measuring and transferring an inert solvent until you are confident that you can handle more hazardous reagents safely. You will often need to hold a syringe full of liquid in one hand while performing another operation with the other hand. One way to do this is as follows:

1. Fill the syringe with the required volume of liquid using both hands.
2. Place the syringe against the palm of one hand using the third, fourth, and fifth fingers of the same hand to grip the barrel.
3. Carefully place the forefinger around the plunger and against the end of the barrel to hold the plunger firmly in place.
4. The liquid can now be delivered by pressing the plunger down using the thumb, and the other hand is completely free.

This method should be practiced using solvent until your technique is good enough to confidently transfer liquids without spillage.

To prepare a syringe for use with moisture-sensitive reagents, first select a clean, appropriately sized syringe and needle. Dry both of these in an oven and allow them to cool in a desiccator. Before using the syringe, it should be purged with argon (or nitrogen), and this is conveniently done by "syringing" the gas via a septum inlet attached to the inert gas system (Figure 6.14). This should be repeated several times. After purging with argon (or nitrogen), the syringe can be kept for a short time before use if the tip of the needle is inserted into a small rubber stopper to prevent air from entering the needle.

Before attempting to syringe liquid from a bottle that is under an inert atmosphere, make sure you have an effective inert gas system with a needle outlet; a dry syringe (and preferably a spare) prepared as described earlier; and a receptacle, *also under inert atmosphere*, into which you are going to transfer the liquid (Figure 6.15). The procedure outlined in the following should be practiced with solvent first if you have not done it before:

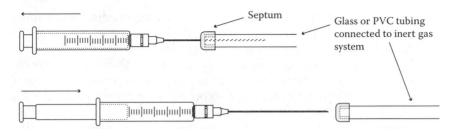

Figure 6.14 Flushing a syringe with inert gas.

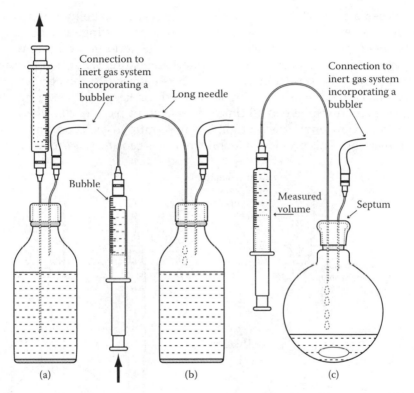

Figure 6.15 Transferring an air- or moisture-sensitive liquid by syringe: (a) fill the syringe with inert gas, (b) force excess reagent and gas bubbles back into the bottle, and (c) deliver measured volume of reagent.

1. Pressurize the reagent container using an inert gas line with a syringe needle attached.
2. Carefully insert the needle of the syringe through the septum, holding the syringe plunger in. Then *very slowly* fill the syringe by pulling the plunger up (Figure 6.15a). You should never fill the syringe to more than about two-thirds capacity.

3. Tip the syringe upside down, bending the long needle. Slowly force any excess reagent and gas bubbles back into the bottle until the required volume is indicated (Figure 6.15b).

4. Holding the syringe barrel carefully in place with one hand, slowly withdraw the needle from the septum using your other hand to control it. Then, avoiding prolonged exposure to the atmosphere, *quickly* insert the needle into the receptor.

5. Slowly deliver the measured volume into the receiver flask, which should itself be connected to an inert gas system.

When a particularly moisture-sensitive reagent is being transported by syringe, even brief exposure of the tip of the syringe needle to the atmosphere may cause problems. For example, tert-butyllithium, which is pyrophoric on contact with atmospheric moisture, will often produce a flame from the syringe needle tip if it is exposed to the atmosphere. Fortunately, there is a very simple method for keeping the syringe tip under inert atmosphere at all times. Take a short piece of dry glass tubing (about 5 cm long), attach a tight fitting septum in each end, and purge with inert gas. This provides a mobile inert gas capsule (Figure 6.16a) that

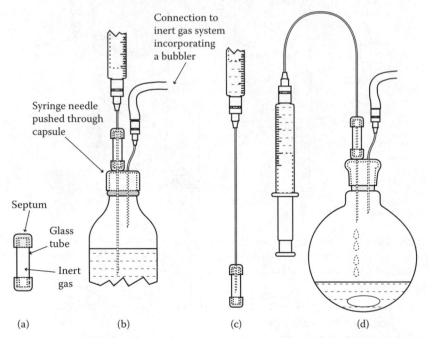

Figure 6.16 Maintaining inert atmosphere around a syringe needle tip: (a) inert capsule used to protect the tip of the syringe, (b) transfer of the reagent to the syringe with inert capsule in place, (c) pull the tip of the needle into the inert capsule, and (d) deliver the reagent.

can be used to protect your syringe tip. The procedure for transferring the liquid is exactly as described previously except for the following points:

1. When drawing the liquid out of the reagent container, the syringe needle is passed through both septa of the inert gas capsule (Figure 6.16b).
2. After filling the syringe, the end of the needle is pulled into the capsule so that the tip is protected (Figure 6.16c).
3. The capsule is pressed against the septum of the receiving flask and the needle is pushed through both septa to deliver the liquid (Figure 6.16d).

When using containers sealed with Mininert valves, a slightly different syringe technique is required. This is because Mininert valves are too narrow for two syringe needles to pass through them. It is therefore impossible for a container fitted with a Mininert valve to be pressurized from an inert gas system while syringing from it. However, because the containers used with these valves are usually very small, a simple alternative procedure can be used:

1. After flushing a syringe with inert gas, fill it with slightly more gas than the quantity of liquid you wish to syringe (a gas-tight syringe is preferred for this procedure).
2. Open the Mininert valve, push the syringe needle through, and push the plunger to pressurize the container with the inert gas from the syringe.
3. Lower the needle below the surface of the liquid and allow the syringe to fill under the pressure that you have introduced (never suck liquid into the syringe as this will draw air into the system).
4. Bend the needle to invert the syringe, displace gas bubbles, and measure the required volume of liquid.
5. Remove the needle from the container and deliver the liquid as usual.

If you need to add 2–50 cm³ of a reagent by syringe slowly over an extended period of time, it is best to use a syringe pump (Figure 4.16). To do this, the syringe is prepared in exactly the same way as for manual addition and then loaded onto the syringe pump. The pump is then set to add the reagent over the desired time and started. Simple syringe pumps only allow you to set the rate of addition, but more sophisticated versions are also available that will allow the syringe pump to be controlled by software linked to an output from a reaction monitoring device. For example, the rate of addition could be set to maintain a particular internal temperature during an exothermic reagent addition.

6.4.5   Handling and weighing solids under inert atmosphere

Manipulation of solid reagents under dry conditions tends to be more awk-ward than handling liquids or solutions. However, there are two distinctly different problems that can be considered. First, there is the use of dry but unreactive solid reagents in reactions under inert conditions, and this will be dealt with in Chapter 9. The other and more significant problem is manipulation of very reactive solid reagents, which may react with mois-ture or oxygen in the atmosphere. The only way to handle solids under completely inert conditions is to use a glove box, which is a sophisticated and relatively expensive piece of equipment. Fortunately, there are rela-tively few reagents that are used routinely in organic synthesis that need to be handled in a completely inert environment. If you do not have access to a glove box and need to handle a particularly reactive solid, a glove bag can be used. This is essentially a large, clear polythene bag, which can be sealed after equipment and reagents have been placed in it. Inlets are pro-vided for electricity cables, gas lines, and so on, and glove inserts allow for the manipulation of the solid and reaction apparatus inside the bag. After placing the reagent container and all the necessary equipment into the bag, it is evacuated and filled with an inert gas three times (a double manifold can be used for this purpose). Once this has been done, all oxygen should have been removed from the interior of the glove bag. At this point, the reactive solid container can be opened and the solid added to the reaction flask. Some skill is required to work within a glove bag, so as usual when working under inert conditions, plan your work very carefully making sure everything you require is in hand before you start. It is also a good idea to practice working in the bag before you attempt the "real thing."

For handling the vast majority of solid reagents used in organic syn-thesis, a glove box/bag is unnecessary, provided you master a few simple techniques and work carefully and quickly. For commonly used moisture-sensitive solids such as alkali earth metals (sodium, lithium, potassium, etc.) and metal hydrides (sodium hydride, potassium hydride, lithium aluminum hydride, etc.), the most commonly encountered problem is accurately weighing them out. We have already shown that measuring liquids under inert atmosphere is relatively simple, and for this reason, many air-sensitive organic solids are sold in solution. If you need to weigh the solid itself without using a glove box/bag, a certain amount of expo-sure to the atmosphere will be inevitable, but this can be minimized, pro-vided you are careful. When carrying out a reaction, it is normally best if you can plan events so that any air-sensitive solid is added to the reaction flask before other reagents or solvent.

Reactive metals (e.g., lithium, sodium, and potassium) are often sup-plied in rod or block form and stored under oil to prevent exposure to atmospheric moisture. To weigh a reactive metal, first make sure you have a pre-dried flask containing an inert atmosphere available. Although the metal will normally be stored under paraffin oil, it may have an oxide or hydroxide layer coating it. Both the oil and the coating need to be removed before weighing the metal. This can be done using the following sequence (Figure 6.17). *Remember that most reactive metals are pyrophoric when they come into contact with moisture.*

1. Place the metal into a beaker, covering it with oil, and then cut some of it into small pieces with a scalpel, removing any coating, leaving a shiny surface exposed. Take care to ensure that the newly exposed shiny surfaces are kept below the surface of the oil. *Heavily coated potassium has been known to detonate on cutting and should be discarded.*
2. Using a pair of forceps, quickly transfer the metal chunks to a second beaker containing *dry* petroleum ether to wash the oil away.
3. Quickly transfer the chunks of metal, allowing the solvent to evaporate very briefly, to a pre-weighed beaker of dry oil. You may find it convenient to have this on a balance so that the additional weight of each chunk of metal can be measured immediately.
4. Once you have transferred the required quantity of metal to the pre-weighed beaker, remove the metal chunks, washing again in the *dry* petroleum ether, and quickly add them to the reaction flask while maintaining a positive pressure of inert gas within the flask.
5. When weighing is complete, add an alcohol (e.g., ethanol) to the beakers used for the weighing procedure to decompose any traces of reactive metal that may be left.

Finely divided metals (e.g., sodium sand) and metal hydrides (e.g., sodium hydride and potassium hydride) are usually supplied as

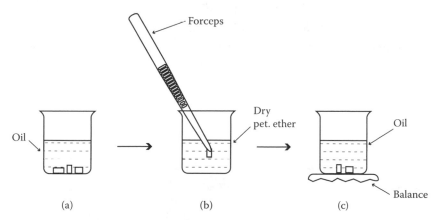

Figure 6.17 Weighing a moisture-sensitive metal. (a) Clean metal cut, (b) wash, and (c) wash into beaker of oil and then rewash.

dispersions in dry paraffin oil. The supplier will specify the amount of metal or metal hydride in the dispersion. However, in some cases (e.g., potassium hydride), the reactive solid may settle in the container leaving a clear layer of oil above. It is important to make sure that the dispersion is evenly distributed before weighing, so if in doubt, shake the sealed container vigorously to ensure this is the case. While dispersed in the oil, the reagents are moderately stable and can be weighed out quickly in the atmosphere.

For some experiments, the dispersion can be used without removing the oil. To do this, simply weigh the dispersion into a pre-dried reaction flask and place it under inert atmosphere by sequentially evacuating and then purging the flask with an inert gas. A three-way Quickfit tap connected to a double manifold is ideal for this purpose.

In most cases, you will need to remove the oil, as this will likely contaminate the final product. There are various techniques for removing the oil from a metal dispersion. The simplest method (Figure 6.18) is as follows:

1. Weigh the dispersion in a flask (do not forget to take the oil into account) and place it under inert atmosphere. Use a double manifold connected by a three-way Quickfit tap if possible (Figure 6.18a).
2. While maintaining a rapid flow of inert gas through the bubbler of the inert gas system, open the three-way tap and add some dry petroleum ether using a syringe. Swirl the flask to dissolve the oil and then let it stand until the solid has settled at the bottom (Figure 6.18b).

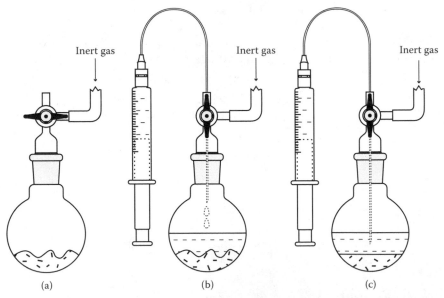

(a)   (b)   (c)

Figure 6.18 Removing oil from a metal dispersion (small-scale): (a) connect to manifold, (b) wash with petroleum ether, and (c) remove the petroleum ether.

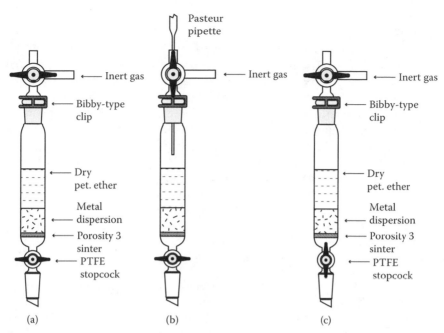

Figure 6.19 Removing oil from a metal dispersion (large-scale): (a) place under a positive pressure of inert gas, (b) add petroleum ether, and (c) force petroleum ether through the dispersion.

3. Draw off the petroleum ether carefully using a syringe, ensuring that the solid remains behind (Figure 6.18c). Add the petroleum ether dropwise into a beaker of *iso*-propanol to ensure that any fine particles of the reactive solid are quenched.
4. Repeat steps 2 and 3 two more times.
5. The flask can then be evacuated to remove the last traces of petroleum ether, purged with inert gas, and then reweighed to determine the exact quantity of solid in the flask. With the flask reconnected to the inert gas system by a three-way tap, reaction solvents and other reagents can then be added and the reaction performed.

If you need to separate the oil from a quantity of metal dispersion without placing it directly into a reaction flask, the piece of apparatus shown in Figure 6.19 is very useful. This essentially allows you do to a filtration under inert atmosphere. It is used as follows:

1. After drying the filtration apparatus and cooling under a stream of inert gas, quickly weigh into it slightly more than the required quantity of the dispersion and then attach a Quickfit three-way PTFE tap, leading to a double manifold, to the top.

2. Evacuate and place under a positive pressure of inert gas (Figure 6.19a).
3. Turn the three-way tap so that inert gas is directed out of the vertical inlet as well as into the funnel. Using a Pasteur pipette, add *dry* petroleum ether via the three-way tap (Figure 6.19b).
4. Remove the pipette and close the three-way tap (Figure 6.19a).
5. Open the stopcock at the bottom of the funnel and allow the positive pressure of inert gas to push the petroleum ether through the sinter (Figure 6.19c).
6. Before the level of the solvent reaches the top of the dispersion, close the bottom stopcock and turn the three-way tap so that argon is directed out of the vertical inlet as well as into the funnel. Then, add more solvent through the vent using a Pasteur pipette (Figure 6.19b).
7. Repeat the washing steps until you are sure that all the oil has been washed away.
8. Then, close the top and bottom stopcocks.

This procedure will leave you with finely divided solid under an inert atmosphere, which can be kept for a few hours in the sealed apparatus without deterioration. The fine powder will be very reactive and therefore great care will be required when handling it.

If you need to weigh reactive solids that are not stored under oil and that are sufficiently stable to be manipulated outside a glove box (e.g., lithium aluminum hydride), then the procedure described in the following can be used. As always, plan ahead when handling reactive materials and make sure you have a dry vessel, filled with an inert atmosphere, into which the solid is to be weighed.

1. Connect a large funnel to an argon gas supply and turn on the gas flow. Then, position the funnel over an argon-filled vessel that you are going to weigh the solid into (Figure 6.20). *Note: Argon is essential for this procedure because it is denser than air and so will provide a blanket of inert gas over the flask.*
2. Position the container holding the reactive solid under the argon stream and quickly weigh the required amount into the flask. **When weighing any finely powdered reactive solid, make sure that the gas flow does cause the powder to disperse through the air as this could lead to a fire.**
3. Place a septum in the top of the flask and connect to an inert gas supply (preferably a double manifold), evacuate very carefully to avoid the power being disturbed, and then refill with inert gas.
4. Carefully neutralize remaining traces of the powder on the spatula using an alcohol.

This inverted funnel technique reduces contact of the reactive solid with moist air and thus lowers the chances of deterioration and also the risk of fire.

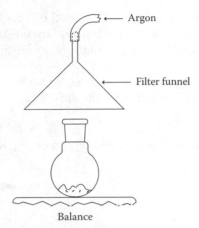

Argon

Filter funnel

Balance

Figure 6.20 Using an inverted filter funnel to provide an argon blanket.

6.5   Preparation and titration of simple organometallic reagents and lithium amide bases

Grignard reagents and organolithium reagents are commonly used as nucleo-philes and bases in organic synthesis. Bulky lithiated amines (lithium amides) are also widely used as strong bases. Some of these reagents are commercially available, but most have to be prepared. Preparation of such reagents is not difficult but a certain amount of care needs to be taken because these reagents are highly reactive toward moisture. In this section, we outline representative methods for preparing the most common organometallic reagents.

6.5.1   General considerations

The most important practical consideration when preparing organometal-lic reagents is that water must be rigorously excluded at all times. Most organometallic reagents cannot be formed in the presence of moisture and once formed they **normally react violently with water**. Make sure you are familiar with the principles outlined in Sections 6.3 and 6.4 and in Chapter 9 before attempting to prepare these organometallic reagents. We recommend that you follow the following general points when preparing any organometallic reagent:

1. All glassware used must be dried rigorously and filled with an inert atmosphere.
2. Coil condensers (Figure 9.26) should be used if possible. These are less prone to problems caused by water condensing on the outside of the condenser, then running down onto the Quickfit joint and seeping into the reaction flask.

3. Quickfit joints on reaction flasks should be lined with PTFE sleeves. This is especially important when fine metal dispersions are used, which may otherwise seep through the joints and cause a potential fire hazard.
4. The outside of all joints should be sealed with PTFE tape to ensure that water and air cannot enter the apparatus.

Typical apparatus setups are shown in Figure 6.21. These can be used for the preparation of most common organometallic reagents.

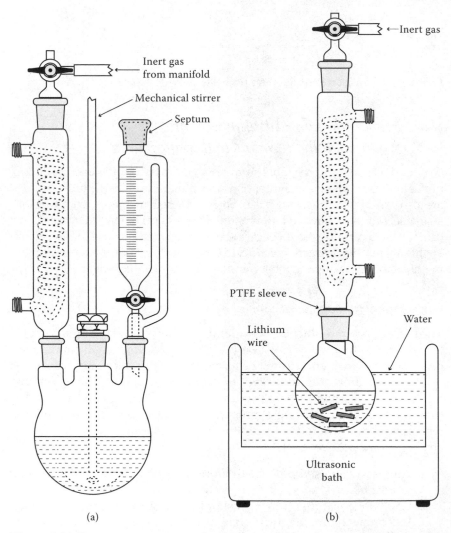

Figure 6.21 Two apparatus setups for the preparation of organometallics: using (a) a mechanical stirrer and (b) an ultrasonic bath.

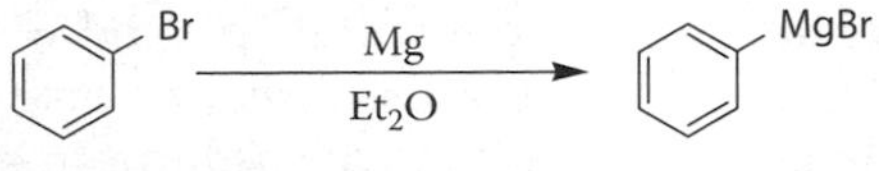

Figure 6.22 Formation of a Grignard reagent.

6.5.2 Preparation of Grignard reagents (e.g., phenylmagnesium bromide)

The apparatus shown in Figure 6.21a, incorporating either a mechanical or magnetic stirrer, can be used for preparation of Grignard reagents (Figure 6.22). All the glassware must be thoroughly dried before use. Dry magnesium turnings (9.72 g, 0.40 mol) are placed in the flask and the system purged with an inert gas (nitrogen or argon). A solution of bromobenzene (45.3 cm^3, 67.5 g, 0.43 mol) in dry diethyl ether (125 cm^3) is then placed in the pressure-equalized addition funnel. A small quantity of the bromobenzene solution (5–10 cm^3) is added to the magnesium to initiate Grignard formation. The formation of the Grignard reagent is indicated by the solution turning cloudy and the production of heat (the diethyl ether should boil gently). If the reaction does not start, add a small crystal of iodine or a few drops of 1,2-dibromoethane and then warm the mixture gently with a heat gun. Once the initial reaction has subsided, the remainder of the bromobenzene solution is added at a rate that maintains gentle reflux. Once the addition is complete, add a small quantity of dry ether (5–10 cm^3) to the addition funnel to rinse any residual bromobenzene into the flask and then stir until all the magnesium has reacted.

The solution produced in this way will contain approximately 0.4 mol of Grignard reagent. It can be reacted in situ by subsequent addition of an electrophile from the addition funnel, or it can be transferred to a storage container by cannulation as described in Section 6.4. Solutions of alkyl and aryl (but not allyl) Grignard reagents can usually be stored for prolonged periods under an inert atmosphere.

Under dry conditions, most alkyl, aryl, and vinyl bromides will react spontaneously with magnesium, as outlined earlier. However, problems initiating the Grignard formation are occasionally encountered. Alkenyl bromides and alkyl bromides possessing proximate ether or acetal groups tend to be the most problematic. Many methods have been recommended for activating the magnesium in such instances.[4] The method we have found to be most straightforward is to stir the magnesium turnings under nitrogen for 10–12 hours before use.

6.5.3 Titration of Grignard reagents

Resublimed iodine (ca. 250 mg, weighed accurately) is placed in a dry 10-cm^3 round-bottom flask or small Pyrex test tube fitted with a magnetic stirrer bar and septum.[5] The system is then purged with argon (or nitrogen) and kept under a positive inert gas pressure, using a needle adapter from an inert gas line or balloon. Anhydrous lithium chloride (4 cm³ of a 0.5 M solution in THF) is then added via a syringe. The mixture is stirred until the iodine completely dissolves and then cooled to 0°C using an ice bath. The solution of Grignard reagent is then added dropwise using an accurate 1-cm³ syringe, until the brown color disappears. At this point, the volume of Grignard solution added is noted. The molarity of the solution can calculated using the following equation:

$$\text{Molarity of Grignard solution} = \frac{\text{mg of iodine used} \times 3.94}{\mu\text{l of Grignard required}}$$

To obtain an accurate titer, it is necessary to carry out this titration at least three times and calculate the average of the results obtained.

6.5.4 *Preparation of organolithium reagents (e.g., n-butyllithium)*

The apparatus shown in Figure 6.21b can be used for the preparation of organolithium reagents (Figure 6.23) on a small scale[6] (for larger-scale work, an addition funnel can be incorporated as in Figure 6.21a). First, make sure the apparatus has been thoroughly dried and purged with argon (or nitrogen) and then place freshly cut lithium wire (5 g, 0.72 mol) in the reaction flask. Add dry hexanes (80 cm³) through the three-way tap, via syringe. Sonicate the reaction mixture using an ultrasonic bath and then add a solution of freshly distilled *n*-butyl chloride (37 cm³, 32.8 g, 0.35 mol) in dry hexanes (50 cm³), dropwise via syringe, over a period of about 10 minutes. n-Butyllithium formation should start after a short induction period (ca. 5–10 minutes), causing the reaction mixture to warm and a purple precipitate to form. Sonicate the mixture for a further 3 hours and then leave it to settle. Once the precipitate has settled, quickly replace the condenser by a septum and carefully cannulate the clear organolithium solution into a clean, dry storage flask, as described in Section 6.4. If you have access to a centrifuge, it is a good idea to use a centrifuge flask fitted with a septum as a storage flask. It can be centrifuged to ensure that any remaining precipitate is deposited as a hard layer at the bottom of the flask. Clear solutions of the reagent can then be taken from the flask by syringe, as required. As long as the solution is protected from atmospheric moisture and oxygen, it can be stored for a prolonged period before use. The molarity of the n-butyllithium solution produced in this

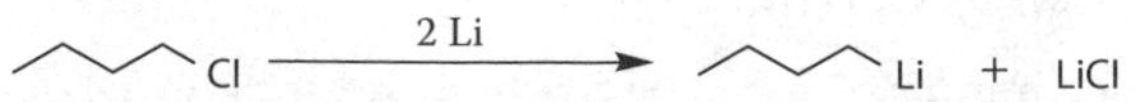

Figure 6.23 Formation of an organolithium reagent.

way should be about 1.8 M, but this should *always* be checked by titration (Section 6.5.5) before use.

The condition of the lithium used in this procedure can be crucial to the success of the reaction. Lithium with a high sodium content is the most reactive and the surface of the metal should be cleaned thoroughly before use. Lithium reagents derived from vinyl bromides can be particularly difficult to prepare. In these instances, it is generally better to use a lithium dispersion that has a high surface area and so tends to be more reactive than chunks of lithium metal. Lithium dispersions are usually supplied in oil, and methods for removing this are given in Section 6.4.5. The resulting metal powder is highly reactive and should be handled very carefully. An alternative method for preparing alkenyllithium and aryllithium reagents on a small scale is to react the appropriate bromide with 2 equivalents of tert-butyllithium. An equivalent of lithium halide is produced when organolithium reagents are prepared from organohalides. This is sometimes undesirable and in such instances the lithium reagent needs to be prepared by a longer route via the appropriate organotin reagent.

6.5.5   Titration of organolithium reagents (e.g., n-butyllithium)

1,3-Diphenyl-2-propanone *p*-toluenesulfonylhydrazone (ca. 200 mg, weighed accurately) is placed in a dry 10-cm^3 round-bottom flask or small Pyrex test tube fitted with a magnetic stirrer bar and septum[7] (Figure 6.24). The system is then purged with argon (or nitrogen) and kept under a positive inert gas pressure, using a needle adapter from an inert gas line or balloon. Dry tetrahydrofuran (4 cm^3) is then added via syringe. The reaction mixture is stirred rapidly to dissolve the hydrazone. The n-butyllithium solution is then added dropwise using an accurate 1-cm^3 syringe until the colorless solution reaches an orange-red color that *just* persists. At this point, the volume of n-butyllithium solution added is recorded, and from this, the molarity of the solution can be calculated using the following equation:

$$\text{Molarity of } n\text{-BuLi solution} = \frac{\text{mg of hydrazone used} \times 2.64}{\mu\text{l of } n\text{-BuLi required}}$$

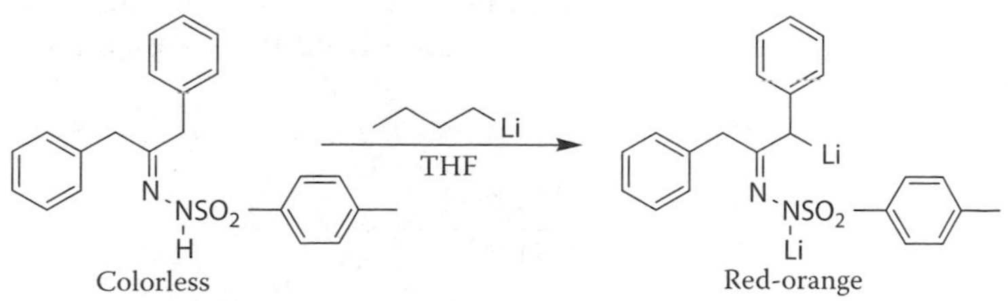

Figure 6.24 Titration of an organolithium reagent.

To obtain an accurate titer, it is necessary to carry out this titration at least three times and calculate the average of the results obtained.

6.5.6   Preparation of lithium amide bases (e.g., lithium diisopropylamide)

Lithium amide reagents are very sensitive to moisture and they must be prepared carefully if they are to be used successfully.[8] A few simple precautions should ensure that the preparation is successful (Figure 6.25):

1. Titrate the n-butyllithium reagent immediately before using it to prepare a lithium amide.
2. Only use diisopropylamine that has been freshly distilled from an effective drying agent (see Section 6.3.1).
3. Make sure the apparatus has been dried rigorously and purged with inert gas before you start.

Make sure you are familiar with the principles of how to perform low temperature reactions before you start (see Section 9.4.1). The reaction flask setup shown in Figures 9.10 and 9.13 can be used for the preparation of lithium amide bases.

To prepare lithium diisopropylamide (LDA), connect a dry 50-cm^3 round-bottom flask containing a magnetic stirrer bar to an inert gas system. Maintaining a positive pressure of inert gas, add dry diisopropylamine (1.4 cm^3, 1.0 g, 10 mmol) via syringe and then add dry tetrahydrofuran (10 cm^3). Cool the mixture to −10°C in an ice–acetone bath. Once the flask has cooled down, add n-butyllithium solution (5.6 cm^3 of a 1.8 M solution in hexanes, 10 mmol), dropwise via syringe, under a positive pressure of inert gas. Once all the n-butyllithium has been added, allow the solution to warm to 0°C and stir for 15 minutes. The LDA solution is now ready and can be used immediately in situ or it can be allowed to warm to room temperature, sealed with a septum, and stored under an inert atmosphere for future use.

For most purposes, it is reasonable to assume quantitative formation of LDA using this method; consequently, if the n-butyllithium solution has been accurately titrated, the amount of LDA formed should be known, but if problems are encountered, there are procedures available to check the titer of the LDA.[9]

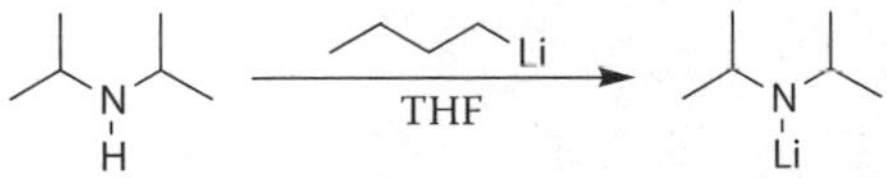

Figure 6.25 Preparation of LDA.

6.6   Preparation of diazomethane

Diazomethane is an extremely useful reagent but is hazardous to prepare and use.[10] The most common use of diazomethane involves reaction with carboxylic acids, which produces methyl esters under very mild conditions. This and other reactions of the reagent have been well reviewed.[10] However, *there are serious hazards associated with its preparation and use.* Consequently, it is essential that you follow appropriate safety measures to minimize the risks involved.

6.6.1   Safety measures

There are two types of hazard associated with diazomethane: toxicity and detonation. Since the reagent is a gas, it can easily be inhaled causing serious lung disorders and a cancer risk. The precursors to diazomethane are also highly toxic and should also be treated with extreme care. Diazomethane has been known to explode. Known detonation initiators include rough glass surfaces, alkali metals, certain drying agents (e.g., calcium sulfate), and strong light.

These risks can be minimized by adhering to the following simple safety measures:

1. *Only prepare diazomethane in small quantities.*
2. *Diazomethane should always be prepared and used in an efficient fume hood, behind a safety blast shield.*
3. *The detonation risk is chiefly associated with concentrated solutions or neat diazomethane. This risk can be substantially reduced by only handling diazomethane in dilute solution.*
4. *Never use diazomethane with ground glass joints or glass apparatus with rough or broken edges and do not use anti-bumping chips.*
5. *If a dry solution is required, use potassium hydroxide as the drying agent.*
6. *Avoid strong light.*
7. *Do not store diazomethane, use it immediately, and neutralize any excess reagent by adding an excess of acetic acid.*

6.6.2   Preparation of diazomethane (a dilute ethereal solution)

"Clearfit" apparatus can be used to prepare diazomethane, but a cheaper alternative is a simple one-piece distillation apparatus, as shown in Figure 6.26, which can be made by a glassblower. It is essential that all the glass surfaces and edges in this apparatus are smooth. If any chips or scratches occur, they must be repaired before use. This setup can be used to prepare between 3 mmol and 17 mmol of diazomethane.

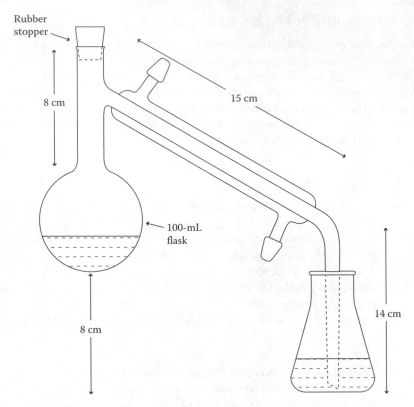

Figure 6.26 Apparatus for preparing diazomethane solution.

Method 1 (3-mmol scale):

Dissolve potassium hydroxide (1 g) in water (2 cm³), dilute with ethanol (8 cm³), and add to the reaction flask. Cool the flask in an ice
bath, then slowly add a solution of Diazald (*N*-methyl-*N*-nitroso-
p-toluenesufonamide) (1 g) in diethyl ether (15 cm³), using a Pasteur
pipette with a flame-polished tip. When all the Diazald has been
added, stopper the neck of the flask, put the ice bath under the
receiver, and place a warm water bath (about 65°C) under the reaction flask. The yellow diazomethane–ether mixture will start to distill out of the flask and collect in the receiver. Continue distilling
until the yellow color has disappeared from the reaction flask, adding a further 5 cm³ of diethyl ether if necessary. The yellow distillate
contains about 3.3 mmol of diazomethane.

Method 2 (17-mmol scale):

To produce diazomethane (16.6 mmol), use Diazald (5 g), potassium
hydroxide (5 g), water (8 cm³), ethanol (10 cm³), and diethyl ether
(50 cm³). The procedure is similar, but on this larger scale, add the

Diazald solution slowly to the warm potassium hydroxide solution, while distilling off the diazomethane. A separating funnel (with a PTFE stopcock) pushed through a rubber stopper can be used for the addition, but make sure that the dropper at the bottom of the separating funnel has been flame polished to remove any rough edges.

6.6.3 General procedure for esterification of carboxylic acids

To an ice-cooled solution of the carboxylic acid in ether or ethanol, add an ethereal solution of diazomethane slowly, until gas evolution stops and a very pale yellow color (excess diazomethane) remains. Next, add just enough dilute acetic acid in ether to quench the yellow color, then concentrate the mixture under reduced pressure, and purify the product as appropriate. For very acid-sensitive reactants, the acetic acid work-up can be omitted and excess diazomethane removed by bubbling nitrogen through the reaction mixture (in an efficient fume cupboard) until all the yellow color has disappeared.

6.6.4 Titration of diazomethane solutions

Add an aliquot of diazomethane solution to an accurately weighed, excess quantity of benzoic acid in ether. Dilute the mixture with ether and titrate with a standard alkali solution to calculate the quantity of acid remaining.

References

1. D.D. Perrin and W.L.F. Armarego, *Purification of Laboratory Chemicals*, 3rd ed., Pergamon Press, Oxford, 1988.
2. L.F. Fieser and M. Fieser, *Reagents for Organic Synthesis*, Vols. 1–13, Wiley, New York.
3. *Organic Synthesis*, Col. Vols. 1–6, Wiley, New York.
4. Y.H. Lai, *Synthesis*, 1981, 13, 585.
5. A.G. Massey and R.E. Humphries, *Aldrichimica Acta*, 1989, **22**, 31.
6. A. Krasovskiy and P. Knochel, *Synthesis*, 2006, 38, 890.
7. H. Gilman, J.A. Beel, C.G. Brannen, M.W. Bullock, G.E. Dunn, and L.S. Miller, *J. Am. Chem. Soc.*, 1949, **71**, 1499.
8. J. Suffert, *J. Org. Chem.*, 1989, **54**, 509.
9. H.O. House, D.S. Crumrine, A.Y. Teranishi, and H.D. Olmstead, *J. Am. Chem. Soc.*, 1973, **95**, 3310.
10. R.E. Ireland and R.S. Meissner, *J. Org. Chem.*, 1991, **56**, 4566.
11. T.H. Black, *Aldrichimica Acta*, 1983, **16**, 3.

chapter seven

Gases

7.1 Introduction

Many organic reactions require the use of gases, either inert gases that are used to protect a reaction or reagent gases that participate in the reaction. Special experimental techniques are required for handling gases, and this chapter contains a summary of methods for the preparation, handling, and measurement of the most commonly encountered gases.

It must be emphasized at the outset that many gases are very hazardous, either because they are toxic or because they are supplied in cylinders that contain compressed gas at very high pressures, and, as a result, when manipulating gases particular attention must be paid to safe practice as outlined in this chapter.

7.2 Use of gas cylinders

A large number of gases are commercially available, and details about individual gases are provided in Sections 7.5 and 7.6. The purpose of this section is to describe recommended methods for handling the containers in which these gases are supplied.[1,2]

Most gases are supplied in pressurized metal cylinders of sizes ranging from 50 cm × 350 cm lecture bottles to 0.25 m × 1.5 m gas cylinders. Some gases with higher boiling points are supplied at lower pressures in relatively light metal cylinders. The fittings on the cylinders may vary depending on the supplier and the gas; hence, it is essential to comply with the supplier's instructions regarding fittings and accessories.

Most cylinders are fitted with a head unit containing an on-off valve, and an outlet screw fitting where a pressure regulator must be attached (Figure 7.1). *This head unit should never be tampered with*. It is a weak point in the cylinder and can be dislodged or damaged if the cylinder is dropped. Apart from releasing a potentially dangerous gas, any damage to the head unit can cause the highly pressurized gas to vent uncontrollably, converting the cylinder into an extremely dangerous missile. *For this reason, cylinders and lecture bottles should never be allowed to stand unsupported*. They should always be securely clamped to a bench or a wall. If they must be moved frequently, they should be supported in

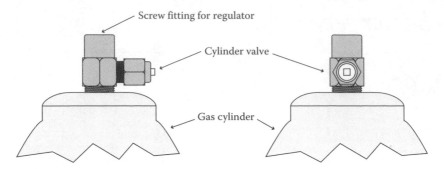

Figure 7.1 Gas cylinder head unit.

a sturdy metal frame or in a trolley designed for this purpose. Cylinders should only be moved in purpose-designed trolleys and should always be treated with great care.

Cylinders are generally pressurized to 175–200 atm, and the on-off valve provides no more control than its name suggests. So cylinders must always be fitted with a pressure regulator before use. A two-stage pressure regulator (Figure 7.2) is normally used with large gas cylinders. Pressure regulators provide a constant outlet pressure that can be adjusted to suit a particular application. If you need precise control over outlet pressure, it is a good idea to fit an additional needle valve to the regulator outlet. A very useful accessory is a multiway needle valve outlet (Figure 7.2) that allows more than one apparatus to be connected to the cylinder at one time.

7.2.1   *Fitting and using a pressure regulator on a gas cylinder*

To fit a pressure regulator to a gas cylinder, you need to do the following:

1. Ensure that you have the correct regulator for the gas cylinder in question. Different gas cylinders have different fittings, and it is essential to ensure that the gas does not come in contact with incompatible materials. So the appropriate regulator, fitted with the correct threaded connector, must be used. Gas suppliers should provide detailed information on which regulators are compatible with which cylinder. For safety reasons, it is best to choose a regulator that has a fairly low delivery pressure ($\leq$0–50 psi) unless a high output pressure is specifically required (1 atm = 1.01 bar = 14.5 psi [lb·f/in.^{2}] = 1.03 kg·f/cm^2 = 760 mmHg).
2. Remove the protective plastic cap from the cylinder fitting and ensure that the fitting is clean and dry. *Never use grease or PTFE tape on cylinder fittings*.

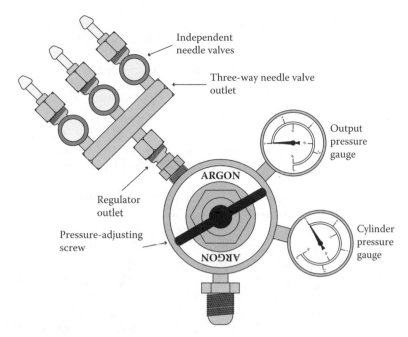

Figure 7.2 Gas cylinder regulator plus three-way needle valve outlet.

3. Screw the fitting onto the cylinder and tighten firmly. A poor fit probably means that the regulator is not of the correct type (note that some cylinders have left-hand threads). *Never try to force the fitting.*
4. Test for leaks by applying a little dilute soapy water or a commercial leak detection solution around the joint. If a leak is present, you will see bubbles forming in the liquid around the joint.

Once you are certain that the regulator is attached correctly, you are ready to use the cylinder. The procedure for the controlled delivery of a gas is as follows:

1. Turn the delivery pressure–adjusting screw anticlockwise until it rotates freely to ensure that the regulator is closed. Also, close any needle valves attached to the regulator.
2. Open the cylinder valve by slowly turning the valve on the cylinder head anticlockwise. You will require a cylinder key to do this. Turn the cylinder key until the cylinder pressure gauge shows the tank pressure. Do not open the valve any further.
3. Check for leaks by applying a little dilute soapy water or a commercial leak detection solution around the joint between the regulator and the cylinder head.

4. If no leaks are present, turn the delivery pressure–adjusting screw clockwise until the required output pressure (typically 10 psi) is registered on the outlet pressure gauge.
5. Attach your gas line to the apparatus; then open the needle valve slowly to obtain the desired flow rate.

When the gas is no longer required, its flow should be shut off and the cylinder should be closed by closing the regulator (rotate the pressure-adjusting screw anticlockwise) and the needle valve and then the cylinder valve. The cylinder valve should always be left shut when the gas is not in use.

Lecture bottles should also be fitted with a regulator and a needle valve, as specified by the manufacturer. There are two main types of regulators for lecture bottles, one for corrosive gases and one for noncorrosive gases. It is essential that you use the correct one. The procedures for using lecture bottles are the same as those for using larger cylinders.

Some gases are supplied in liquid form at relatively low pressures, and these generally require the use of only a flow control valve. The valves on these cylinders are generally fitted with a handwheel. Again, you should consult the supplier's technical data for information on correct fittings and handling procedures.

7.3   Handling gases

This section is concerned with general procedures for handling gases and includes a discussion on apparatus and techniques used for adding gases to reaction flasks. The preparation and scrubbing of some common reagent gases are described in Section 7.6. Methods for the use of inert gases to protect air-sensitive reactions are detailed in Chapter 9. Hydrogenation, ozonolysis, and liquid ammonia reactions are described in Chapter 14.

Two important principles must be borne in mind when carrying out a reaction involving a gas:

1. A pressure release point, such as an oil bubbler or a mercury bubbler, must be attached to the apparatus in order to prevent the possibility of dangerous pressure buildup.
2. Pressure fluctuations may result in the reaction mixture being sucked out of the reaction flask, so valves or traps must be placed in the gas line to suck the mixture back.

Reactions involving toxic and malodorous gases must always be carried out in an efficient fumehood with appropriate scrubbers in place to prevent the escape of gas.

A typical arrangement for the addition of a gas to a reaction mixture is shown in Figure 7.3. The gas is either supplied from a cylinder or prepared in another vessel and is supplied at a steady flow rate. An appropriately sized trap, such as a Dreschel bottle or an empty bubbler, needs to be placed between the gas supply and the reaction flask. This is needed to prevent the reaction contents from being sucked back into the gas source if a pressure reversal occurs. The gas can be bubbled into the solution via a Pasteur pipette or a glass tube fitted with a wide-bore glass frit. A frit will give a better dispersion of the gas, but care should be taken to ensure that the frit does not become blocked by a solid product.

It is important to place an oil bubbler or a mercury bubbler between the gas supply and the trap because if the gas line is blocked the gas can be vented safely. Obviously, the depth of oil or mercury in the bubbler must be large enough to ensure that the gas passes through the reaction mixture rather than venting from the bubbler. Take care to ensure that the liquid in the bubbler is compatible with the gas being used. *Mercury should never be used with ammonia or acetylenes (alkynes)*. With the inclusion of a bubbler as a safety device, it is advisable to put clips on all the ground glass joints, or to wrap them securely with PTFE tape, so that the gas cannot leak from the apparatus.

A bubbler should also be included at the end of the apparatus so that the gas that is not absorbed by the reaction mixture can escape. Small quantities of nontoxic gases can be allowed to vent into an efficient fumehood, but toxic or malodorous gases must be passed through a scrubbing system to ensure that they are not released into the environment. Some of the procedures for scrubbing these gases are described in Section 7.6. The bubbler also provides a means of monitoring the rate of uptake of a

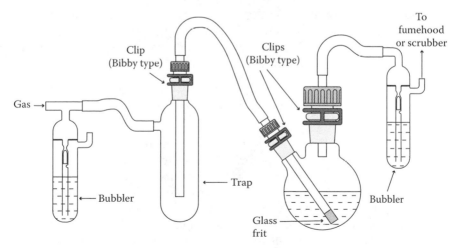

Figure 7.3 Typical arrangement for the addition of a gas to a reaction flask.

gas, and the flow rate should be adjusted so that very little is vented. For relatively insoluble gases or for slow reactions, it may be necessary to stir or shake the reaction flask.

The type of tubing required to connect the various components in an apparatus used in a reaction involving a reactive gas will depend on the nature of the gas in question. Rubber or PVC tubing is convenient for non-corrosive, nontoxic gases such as carbon dioxide, but many reagent gases require the use of glass or chemically resistant plastic tubing. A brief guide to the chemical resistance of some common tubing materials is provided here, and more detailed information on specific products can be obtained from manufacturers; particular caution should be exercised when corrosive and toxic gases are used:

> Polyvinyl chloride (Tygon) tubing: This is flexible with low permeability but poor resistance to organic solvents and acidic gases.
> Polyethylene or polypropylene tubing: This has better resistance to solvents and acids than PVC but is not suitable for halogens.
> Fluorocarbon (e.g., PTFE) tubing: This tends to have poor flexibility but excellent chemical resistance.
> Natural rubber tubing: This has high gas permeability and low chemical resistance, so it should not be used in gas lines.
> Neoprene tubing: This has relatively good resistance to organic solvents and to acids.
> Fluorocarbon rubber (e.g., Viton): This has good chemical resistance but is unsuitable for amines and ammonia.

7.4   Measurement of gases

The addition of an accurately measured quantity of a gas is sometimes necessary, for example, to obtain selective hydrogenation of a diene. Some convenient and reasonably accurate methods for measuring gases are described in Sections 7.4.1 through 7.4.6.

7.4.1   Measurement of a gas using a standardized solution

If a gas is soluble in a suitable solvent and the concentration of the solution can be determined by a simple analytical technique, then accurately measured quantities of the gas can be dispensed by using the appropriate volume of the solution. For example, solutions of hydrogen halides (e.g., HCl) in various solvents can be determined by simple acid/base titration, and solutions of chlorine in carbon tetrachloride can be determined by the addition of excess potassium iodide followed by back titration with sodium thiosulfate. Refer to textbooks on inorganic analysis for details of these methods.

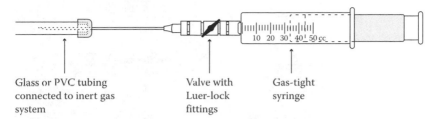

Glass or PVC tubing connected to inert gas system

Valve with Luer-lock fittings

Gas-tight syringe

Figure 7.4 Setup for dispensing gases via a gas-tight syringe.

7.4.2 Measurement of a gas using a gas-tight syringe

Commercially available gas-tight syringes provide the most convenient method for dispensing small volumes of gases. A syringe fitted with an on-off valve and a needle (Figure 7.4) should be repeatedly filled with the gas and emptied to ensure that all air has been removed from the syringe. It should then be filled to the required volume from the gas stream, and the valve should be closed to prevent gas release.

Just before use, the valve needs to be opened and the needle purged by expelling a volume of gas equivalent to the volume of the needle. The needle should then be inserted immediately into the reaction flask. If a small volume of gas is involved (less than 50 cm^3), the reaction vessel can be a sealed system and the gas can be added slowly from the syringe. For larger volumes, the reaction flask should be fitted with a mercury bubbler and the gas should be added at such a rate that it does not vent from the bubbler.

7.4.3 Measurement of a gas using a gas burette

Gas burettes are most commonly used for low-pressure hydrogenations, although they can be used for delivering other gases and provide an easy method of dispensing accurately measured volumes. A simple gas burette design is shown in Figure 7.5 (a more sophisticated version of this design is shown in Chapter 14, Figure 14.1). It is operated as follows:

1. Open the double-oblique tap to the vent (fumehood) and the three-way tap to the burette so that residual air/gas can be vented from the calibrated burette into the fumehood. Raise the leveling bulb to expel most of the residual air/gas from the burette.
2. Turn the double-oblique tap to the gas line so that gas can flow from the supply into the calibrated burette. Turn on the gas supply and slowly fill the burette with the gas, lowering the leveling bulb as the burette fills.

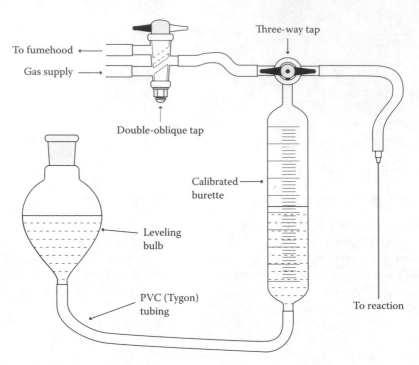

Figure 7.5 Gas burette setup.

3. Repeat steps 1 and 2 twice more to ensure that all the residual air has been expelled from the burette. Fill the burette to approximately 20% more than the required volume, and then close the double-oblique tap.
4. Open the three-way tap to connect the burette to the line with the needle (make sure that the needle is inert to the gas that is being used), and allow the gas to flush out any air from the needle. Then quickly read the volume in the burette as you insert the needle into the reaction flask.
5. Raise the leveling bulb slowly to keep a slight positive pressure of the gas as the reaction proceeds.
6. Once the required volume has been added, close the three-way tap. If the gas is hazardous, make sure you flush the apparatus with an inert gas (via steps 1–4) after use.

If a gas burette is likely to be used with a variety of different gases, then some important safety points should be noted:

1. A logbook should be kept next to the gas burette, in which each user must record their name, the date, and the gas used.
2. The burette should be thoroughly flushed with an inert gas after each use.

3. The tubing may need to be changed for use with different gases, and the liquid in the reservoir may also need to be changed depending on the application (aqueous copper sulfate, mercury, mineral oil, and dibutyl phthalate are often used).

The gas burette method is particularly good for reactions in which the uptake of gas is slow. Gas burettes can also be used for measuring the volume of gas evolved during a reaction.

7.4.4 *Quantitative analysis of hydride solutions using a gas burette*

Another useful application of a gas burette is in the quantitative analysis of hydride solutions. This is achieved by adding the hydride in question to an excess quantity of protic or acidic solvent and measuring the volume of hydrogen evolved.[3]

The procedure for the analysis of $BH_3.THF$ is illustrative:

1. Place 50 cm^3 of a 1:1 mixture of glycerol and water in a 100 cm^3 round-bottom flask containing a magnetic stirrer bar. Seal the flask with a septum and insert into it a needle connected to a graduated gas burette (Figure 7.5) that is filled to a known level with liquid (at least 80% full is ideal).
2. Turn the three way tap so that the burette is only open to the flask; then add a few milliliters of the hydride solution in question to the flask via a syringe to saturate the atmosphere in the system with hydrogen.
3. Adjust the height of the leveling bulb so that the liquid levels in the burette and the reservoir are equal; then note the level of the liquid in the gas burette.
4. Now add an accurately measured volume of the hydride solution to the flask with rapid stirring. Ideally, the amount of hydride added should be sufficient to fill more than half of the gas burette with H_2.
5. When hydrogen evolution ceases, adjust the leveling bulb so that the liquid levels in the burette and the reservoir are equal and then record the change in gas volume.
6. Calculate the molarity using the following equation:

$$\text{Molarity of } H^- \text{ in solution} = \frac{(P_a - P_s) \times (V_h - V_a) \times 273}{V_a \times T \times 22.4 \times 760}$$

In the equation, P_a = atmospheric pressure (in millimeters of Hg); P_s = vapor pressure of the solvent (in millimeters of Hg) in the reservoir at temperature T; V_h = volume of hydrogen evolved (in cubic centimeters); V_a = volume of the hydride solution added (in cubic centimeters); and T = temperature (in kelvin).

7. Repeat the measurements until reproducible results are obtained (this method should give an accuracy better than ±5%).

7.4.5   Measurement of a gas by condensation

Many gases have relatively high boiling points (see Appendix 2), so they can be easily liquefied and measured in the liquid form. To do this, the gas is passed into a vessel that is cooled to a temperature well below the boiling point of the substance, as shown in Figure 7.6. The quantity of condensed gas can be measured by volume or by weight. When the required quantity has been condensed, the line to the gas supply is closed and a reaction flask is attached to the anti-suck-back trap. The cooling bath is allowed to warm to a temperature that allows the liquid to boil gently, forcing the gas into the reaction vessel. The bubbler and the anti-suck-back trap in the system serve the usual functions.

7.4.6   Measurement of a gas using a quantitative reaction

A specific quantity of a gas can be added to a reaction if the gas is prepared using a chemical reaction of known yield. Methods and apparatus for

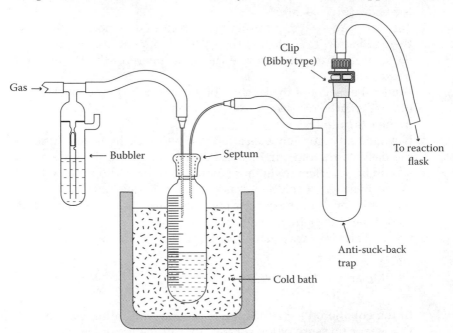

Figure 7.6 Measurement of a gas by condensation.

preparing some commonly used gases are described in Section 7.6.2; further, the elegant gas generator devised by H.C. Brown, which is described in detail elsewhere in the literature,[3] deserves special mention.

7.5   Inert gases

Nitrogen (N_2) and argon (Ar) are by far the most commonly used inert gases, and techniques for conducting reactions under an inert atmosphere are described in Chapter 9. Both N_2 and Ar are available in various grades of purity and, usually, gases of 99.995% purity can be used without further purification. Purification using conventional drying trains is generally counterproductive. An easy method for checking whether your inert gas line is absolutely dry is to attach a small flask containing some titanium tetrachloride to the manifold. Evacuate the flask carefully for a few seconds and then open to the inert gas. If you see white fumes of titanium dioxide, you have a poor batch of gas or a leak in your gas line. Effective gas purification systems are described elsewhere in the literature.[4,5] Although it is more expensive, argon is superior to nitrogen in two important aspects: (1) it is heavier than air and, therefore, it protects the contents of a flask more effectively; and (2) it is completely inert, whereas nitrogen does react with some materials (e.g., lithium metal).

7.6   Reagent gases

Occasionally, it may be necessary to prepare a gas. Methods for the preparation (and scrubbing) of some simple gases are provided in this section. Most of these preparations can be carried out in the apparatus shown in Figure 7.7.

This setup is suitable for generating a gas by reaction of a solution of one reagent with a second reagent whether it is a liquid or a solid. The solution of one reagent is placed in a pressure-equalizing dropping funnel and is added *slowly and cautiously* to the other reagent in the flask with stirring, and cooling if necessary. As the gas is produced, it flows out of the system via the attached tube. This would be attached to a setup such as the one shown in Figures 7.3 and 7.6. The rate of addition of reagents should be adjusted so that the gas is generated at a rate that does not lead to significant pressure buildup in the system. As always, there should be a bubbler attached to the gas generator (Figure 7.7), and there should be a second one at the end of the line (after the reaction flask) to ensure pressure release should a blockage occur in the system. The bubbler at the end of the line also protects the reaction flask from the atmosphere and from any agents (often water or bleach) used in the scrubber. If a toxic gas is generated as the result of the reaction, the apparatus should be flushed thoroughly with an inert gas before dismantling.

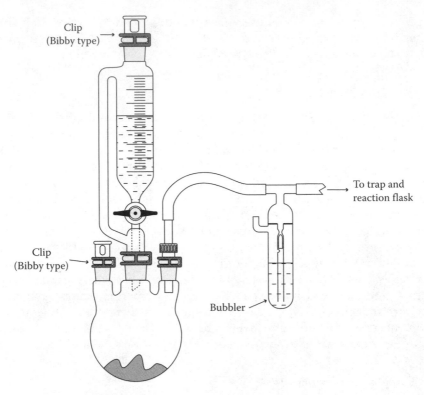

Clip
(Bibby type)

Clip
(Bibby type)

To trap and
reaction flask

Bubbler

Figure 7.7 Gas generator setup.

7.6.1   Gas scrubbers

Gas scrubbers are usually attached to the end of a gas line (after the reaction flask) to ensure that no toxic or malodorous substances are released into the atmosphere. Various types of scrubbers may be employed, but the simplest method of scrubbing relatively small quantities of water-soluble gases is to pass them through water. For malodorous compounds such as thiols, bleach (aqueous sodium hypochlorite) can often be used, but always ensure that the scrubber compound does not react violently with the gas in question. A simple scrubber setup is shown in Figure 7.8. This allows a number of different scrubber solutions to be linked together in parallel. This type of setup is only suitable for scrubbing small quantities of effluent gases, and care should always be taken to ensure that the different scrubber solutions are compatible with one another.

7.6.2   Methods for preparing some commonly used gases

Methods for preparing some common gases are described in this section (see the literature[3] for other methods). The properties of these

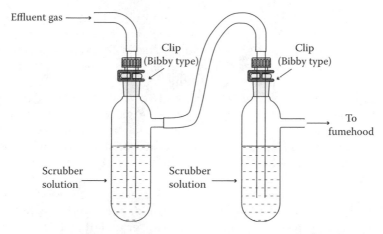

Figure 7.8 Gas scrubber setup.

and other gases are listed in Appendix 2. Many of the following gases are very harmful and should only be handled in an efficient fume cupboard taking care to prevent leaks and to scrub any excess gas:

Acetylene: Acetylene is prepared by the dropwise addition of water to calcium carbide and dried by passing the gas through a column of molecular sieves. Acetylene can explode spontaneously, especially when pressurized. Also, you should never allow acetylene to come into contact with mercury (e.g., in a bubbler). For this reason, when generating this gas, the generator should be cooled, and the gas line should be fitted with an inert oil bubbler so that the pressure cannot rise above 15 psi. Dispose of the gas by venting slowly in an efficient fumehood.

Carbon dioxide: The simplest method of carbon dioxide preparation is to allow solid carbon dioxide (dry ice) to evaporate. The gas can be dried by passing it over molecular sieves. It can also be prepared by the dropwise addition of dilute hydrochloric acid to calcium carbonate. Passage over molecular sieves will remove any aqueous acid in the gas stream.

Carbon monoxide: Carbon monoxide is very toxic by inhalation, its threshold limit value (TLV) is 50 ppm. It is colorless and odorless; so whenever this gas is used, it is essential that a working carbon monoxide detector with an alarm is placed close to the experiment apparatus. This will alert you to any carbon monoxide leak from the apparatus. Carbon monoxide can be generated by the slow addition of anhydrous formic acid to concentrated sulfuric acid at 90°C–100°C (beware, frothing tends to be a problem). It is noted that 1 mL (1 cm³) of formic acid generates 26.6 mmoles of the gas. Carbon monoxide generated in this way is contaminated with small amounts of carbon dioxide and sulfur dioxide, which can be removed by passing the gas over potassium hydroxide pellets or the commercial product Ascarite (sodium hydroxide on silica). Dispose of carbon monoxide by slow venting in an efficient fumehood.

Chlorine: Chlorine is extremely toxic and a powerful irritant (TLV 1 ppm). It also reacts violently with most bases (e.g., NaOH, KOH), so do not use such bases as drying agents or scrubbers. Chlorine is prepared by the slow addition of concentrated hydrochloric acid to potassium permanganate (6.2 cm^3 of acid per gram of permanganate), with occasional shaking. The gas evolution slows as the reaction proceeds, and warming is required to complete the reaction. Theoretically, 1.12 g of chlorine should be formed per gram of permanganate; but in our experience, the yield is substantially lower than this. Excess chlorine can be disposed of by passing through water.

Diazomethane: See Section 6.6.

Hydrogen bromide: Hydrogen bromide is corrosive and toxic (TLV 3 ppm). It can be prepared by the slow addition of bromine to freshly purified tetralin, with stirring. The hydrogen bromide generated in this way will be contaminated with a small amount of bromine, and this can be removed by passing the gas through a trap containing pure dry tetralin. Dispose of hydrogen bromide by passing through water.

Hydrogen chloride: Hydrogen chloride is corrosive and toxic (TLV 5 ppm). It is prepared by adding concentrated sulfuric acid to anhydrous ammonium chloride and dried by passing the gas over molecular sieves. Dispose of hydrogen chloride by passing through water.

References

1. *Safe Under Pressure*, British Oxygen Company, Guilford, Surrey, U.K., 1993.
2. Compressed Gas Association, *Handbook of Compressed Gases*, Reinhold, New York, 1981.
3. H.C. Brown, *Organic Synthesis via Boranes*, Wiley, New York, 1975.
4. D.F. Shriver and M.A. Drezdzon, *The Manipulation of Air-Sensitive Compounds*, 2nd ed., Wiley, New York, 1986.
5. A.L. Wayda and M.Y. Darensbourg, Eds., *Experimental Organometallic Chemistry: A Practicum in Synthesis and Characterization*, American Chemical Society, Washington, 1987.

chapter eight

Vacuum pumps

8.1 Introduction

A variety of tasks in organic chemistry require provision of a vacuum source, and different tasks require different levels of vacuum. Vacuum sources can be loosely divided into three categories:

1. Low vacuum (20–50 mmHg): This is usually sufficient for rotary evaporation of volatile solvents, vacuum filtration, distillation of relatively volatile oils, and similar tasks.
2. Medium vacuum (5–15 mmHg): This is ideal for a variety of tasks including rotary evaporation of less volatile solvents, distillations, and serving double manifold–type inert gas lines.
3. High vacuum (below 1 mmHg): This is most commonly used for removing traces of solvents or moisture from small quantities of material and for the distillation of high-boiling oils. It is also ideal for serving double manifold–type inert gas lines.

8.2 House vacuum systems (low vacuum)

Many large laboratory buildings have a large central vacuum pump that serves as a house vacuum system. There are great variations in the effectiveness of such systems, as the level of vacuum depends on the type of pump employed, the number of users, and the quality of the pipework. A good house vacuum system will typically pull about 50 mmHg and can be useful for driving rotary evaporators, filtrations, and simple vacuum distillations. The advantage of this system is ease of use, but the vacuum can fluctuate considerably depending on how many people are using the system.

8.3 Medium vacuum pumps

8.3.1 Water aspirators

Water aspirators are widely used. They typically produce a vacuum of about 10–15 mmHg when attached to a cold water supply. The advantages of water aspirators are that they are cheap and simple to operate. The main disadvantages are that they can consume large quantities of water

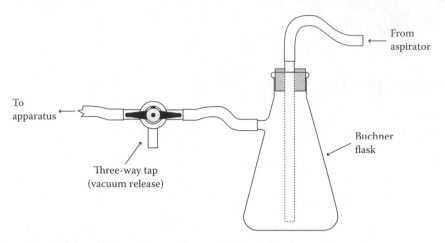

From
aspirator

To
apparatus

Buchner
flask

Three-way tap
(vacuum release)

Figure 8.1 Water trap for use with water aspirators.

and are prone to suck back, which can leave your apparatus full of water. This occurs if the water pressure drops after a vacuum has built up in the apparatus. The same thing will happen if the tap is turned off while the attached apparatus is still under vacuum. For this reason, the tap running the aspirator should always be kept full on, and a vacuum release should be provided by means of a three-way tap. A simple trap (Figure 8.1) should also be incorporated; this will give you some time to disconnect the apparatus if suck back occurs. A water trap will not prevent suck back altogether, and hence it is not advisable to leave an apparatus unattended when it is connected to an aspirator. This type of vacuum source is not suitable for connecting to a manifold-type inert gas line.

8.3.2   *Electric diaphragm pumps*

Oil-free electric diaphragm pumps can be used to provide vacuum levels down to about 5 mmHg. Many of these pumps are PTFE lined, which helps to make them resistant to damage caused by solvent vapors. They are most commonly used for medium pressure distillation and for evacuating rotary evaporators. The problem of suck back is circumvented using a diaphragm pump, and such pumps do not use water. For many tasks, they can also be used instead of high vacuum oil pumps, with the advantage that liquid nitrogen trapping systems are usually not required. For most purposes, they can be used as the vacuum source for double manifold–type inert gas lines.

8.4 High vacuum pumps

8.4.1 Rotary oil pumps

Rotary oil pumps provide reliable vacuums down to about 0.01 mmHg (in an apparatus) and are thus one of the most valuable pieces of equipment in the laboratory. Ideally every research worker should have his or her own high vacuum pump, but in many cases cost prohibits this practice and pumps are shared. If a pump is shared, it is common to have it mounted on a trolley that has all the ancillary devices (traps, manometer, etc.) mounted on it. Alternatively, a shared pump may be fixed in one place but attached to a communal manifold, a distillation setup, or some other piece of apparatus. A good two-stage pump is suitable for most high vacuum requirements in an organic chemistry laboratory.

High vacuum pumps must be fitted with efficient solvent traps to prevent the pump oil from becoming contaminated. The most commonly used traps are cold-finger condensers incorporated into the vacuum line before it enters the pump. It is preferable to use a double-cold-finger condenser system (Figure 8.2), with the traps being cooled by either liquid nitrogen or dry ice/acetone. If a single condenser system is used, liquid

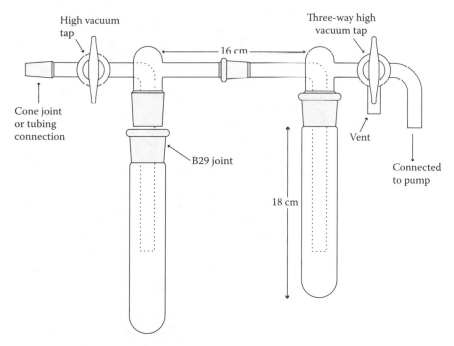

Figure 8.2 Cold-finger condenser solvent trap setup for high vacuum pumps.

nitrogen must be used as the coolant because dry ice/acetone will be ineffective.

Do not immerse the solvent traps in liquid nitrogen unless they are evacuated as this will cause liquid oxygen to condense. The result can be a violent explosion, which is caused by revaporization or oxidation of organic materials, such as solvents

If the trapping system incorporates a cone joint, it can be connected directly to a manifold. Alternatively, a piece of high vacuum tubing can be used to connect the traps to a distillation apparatus, a double manifold, or any other piece of equipment.

To use the cold-finder condenser setup shown in Figure 8.2, you need to attach your apparatus and a high vacuum pump to the positions indicated. *Always check that the traps are empty before switching the pump on.* Then with the two-way tap closed and the three-way tap open between the pump and the traps but closed to the vent, turn the pump on. Next, immerse the traps in Dewars filled with liquid nitrogen and check the vacuum by connecting a vacuum gauge via the three-way tap. If the vacuum is satisfactory (<1 mm), the two-way tap can be opened to evacuate the apparatus that is connected to it. The vacuum should be checked again to make sure that there are no leaks in the system.

When you want to switch the pump off, always close the two-way tap first. Then turn the three-way tap to vent through the pump (*not into the traps*), and then turn the pump off. Remove the liquid nitrogen Dewars from the traps immediately after switching the pump off and vent air into the traps using the three-way tap. It is very important to vent the pump before turning it off; otherwise, pump oil may be sucked back into the traps.

High vacuum pumps are very expensive items, but they give reliable service for many years if they are treated properly. For organic chemists to work efficiently, their pumps must work properly, and this will be the case only if the pumps are treated with care. The main reason for the deterioration of vacuum pumps is oil contamination, either with volatile organic compounds or with acidic vapors such as HCl or HBr. To avoid this problem, you should be very conscientious about keeping the coolant in traps well topped up, and the traps should be emptied and cleaned regularly. No matter how careful you are, some solvent vapor will always find its way into the pump, and for this reason the oil should be changed regularly, at least once a month for a pump that is used every day.

8.4.2   *Vapor diffusion pumps*

Occasionally, a distillation will require a higher vacuum than that produced by a rotary oil pump. Higher vacuums are normally obtained by employing a mercury or oil vapor diffusion pump, in conjunction with a rotary oil pump. Oil vapor pumps are very commonly used today, and various models that will produce a vacuum of about 1×10^{-5} mmHg are now commercially available. These pumps are only used occasionally, but it is very useful to have one shared among a group or a section.

8.5   Pressure measurement and regulation

It is very important to be able to measure the pressure in a vacuum system, particularly when carrying out a distillation. There is a range of simple and accurate mercury gauges that enable accurate pressure measurements, and some of them are described in this section. However, in most laboratories the use of mercury is restricted on grounds of safety, so mechanical or electronic gauges are often preferred. These are also useful for in-line measurements, and they are particularly valuable when used with rotary evaporators.

When using mercury gauges, it is important to follow appropriate safety precautions. For low vacuum measurements, a simple manometer (Figure 8.3a) is commonly used and the pressure is taken by subtracting the heights of the two mercury levels. For high vacuum measurements, a McLeod gauge (Figure 8.3b), with a range of 1–0.001 mmHg, is commonly used. The gauge is rotated to the vertical position to read the vacuum, but it must be turned back to the horizontal position and back again to the vertical position in order to read any change in pressure.

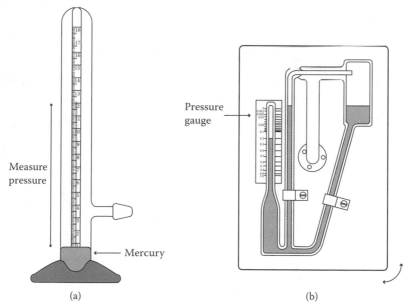

Figure 8.3 Figure showing (a) a mercury manometer and (b) a McLeod gauge.

In many cases, the vacuum required for a distillation will be lower than that generated directly by a vacuum pump (i.e., the required pressure in the system is *higher* than that provided directly by the pump). Consequently, a means of adjusting the vacuum level will be needed. Achieving the required vacuum level is no simple matter because an accurate leak must be provided in order to maintain a constant vacuum. Some electronic vacuum gauges will allow you to do this, and more sophisticated versions even allow you to program the gauge such that the level of vacuum changes over time. However, if they are not available, an accurate needle valve attached to the vent of a high vacuum trap will be suitable for most distillation purposes.

8.5.1   Units of pressure (vacuum) measurement

The different units of pressure measurement often cause confusion, but the way in which they are related to one another is quite simple. The most common unit is millimeters of mercury (mmHg), and 1 mmHg is the same as 1 Torr. Another common scale is millibar (mbar), which is related to atmospheric pressure, 1 bar = 1 atm = 760 mmHg. Thus, 1 mbar = 0.76 mmHg = 0.76 Torr.

chapter nine

Carrying out the reaction

9.1 Introduction

This is perhaps the most important chapter in this book. We would strongly advise anyone new to synthetic organic chemistry research to read through this chapter before starting labwork.

To be certain of the results of a particular experiment, and to ensure reproducibility, it is important that the appropriate precautions are taken and that proper preparations are made. Many common organic reactions involve the use of air- or moisture-sensitive reagents. For these to be successful, it is essential to know how to perform the reaction under rigorously inert, anhydrous conditions. Because of the crucial importance of these reactions, most of this chapter will focus on techniques that are compatible with the use of air- and moisture-sensitive reagents. The same basic techniques are also used for less sensitive reactions, the only difference being that an inert atmosphere may not be required.

Before attempting the reaction, it is important to ensure that you have the necessary glassware, apparatus, reagents, and experimental procedure (including work-up). You should also have a reliable means of monitoring the progress of the reaction (e.g., by TLC, HPLC, and NMR) to determine *exactly* when it is finished. If the product has never been made before, it is important to choose a monitoring system that will be suitable for distinguishing the starting material from the expected product(s). For example, if using TLC, think about whether the polarity is likely to increase or decrease on converting starting material to product and choose a solvent system accordingly. It is also essential that you are familiar with all the hazards associated with the chemicals (including solvents and potential reaction products) and the procedures that are being used. Make certain that you allow yourself plenty of time to complete the reaction or at least to closely monitor it in the early stages if it is likely to be a lengthy experiment. In our experience, many "overnight" reactions reported in the literature are actually complete in a few hours. Leaving an experiment longer than required can only lead to loss of yield or product purity. When planning a reaction, chemists generally spend a considerable amount of time considering the reagents and conditions that should provide good conversion of starting material to product, but an important aspect of the experiment that is often overlooked is how the product will be isolated in pure form. It is important to think about reaction work-up and product isolation from the outset, because an effective work-up can often be dependent on choosing an appropriate reagent and reaction solvent. Chapter 10 provides some general guidance on work-up and isolation. Finally, it is crucial that you accurately record *all* details of the experiment in you laboratory notebook as you go along (see Chapter 3 for more details on how to do this). Remember, it is important to write up experiments that have "failed" just as carefully as those that are successful. If this is done, the reason for the failure might be apparent to you at a later stage as you learn more about the chemistry involved. This detailed information will also be of real value to someone following up your work.

9.2   Reactions with air-sensitive reagents

9.2.1   Introduction

Many commonly used organic reagents are extremely reactive toward water and/or oxygen. Some examples are alkyllithium reagents, Grignard reagents, organoboranes, metal hydrides, organoaluminum compounds, cuprates, titanium tetrachloride, dried solvents, and so on. Some of these reagents are available commercially and others are prepared in situ; in either case they are most conveniently handled in solution. Although extreme care must be taken to exclude air and moisture when using these reagents, you should find them easy to handle once you become familiar with the techniques outlined here. General procedures for handling air-sensitive reagents are described in Chapter 6 and the appropriate sections of that chapter should be consulted in conjunction with this discussion.

9.2.2   Preparing to carry out a reaction under inert conditions

If you already have a permanent inert gas line in the laboratory as suggested in Chapter 4, then carrying out a reaction under inert conditions should be straightforward. However, it is very important that you do not expose the reaction to the atmosphere at any stage. Therefore, it is essential to ensure that everything is in the right place at the right time and under an inert atmosphere. Therefore, before you start, think very carefully about how you intend to carry out each operation of your reaction sequence and list the equipment you will need. Then, make sure that all the equipment has been rigorously dried in advance. It is particularly important to make sure that you have enough syringes available. Indeed, you should always have some spare dry syringes in case one gets jammed, contaminated, or broken.

9.2.3 Drying and assembling glassware

First, make sure that all the glassware is clean. It is also important to check glassware, particularly flasks, for cracks and scratches, especially if the apparatus is going to be evacuated. If you are in any doubt about the integrity of a particular item of glassware, *do not use it*. Either get it repaired by a glassblower or dispose of it. Do not leave it around for someone else to discover its weakness.

Under normal conditions, all glassware will have a thin film of water adsorbed to its surface. This will need to be removed before moisture-sensitive reagents, dried solvents, or starting materials are allowed to come into contact with it. There are two basic methods for doing this. The first method is to heat the glassware in an oven at a temperature above 125°C for at least 6 hours, then allow it to cool in a dessicator, or quickly assemble it while hot and cool under a stream of dry inert gas. The second method is to assemble the glassware, evacuate it by connecting it to a double-manifold/high vacuum pump, and then heat the whole set up with either a Bunsen (flame drying) or a heat gun. Once the water layer has been removed, the two-way tap on the manifold can be switched to fill the apparatus with argon (or nitrogen) while it cools, leaving the apparatus set up and ready for use. The second method can save a good deal of time, but because of the dangers involved in heating glassware under vacuum, it is best reserved for small-scale reactions.

Normal glass syringes can be dried in an oven and then cooled in a desiccator over P_2O_5 or self-indicating silica gel, but it is not advisable to heat microliter syringes; these are best dried under vacuum in a desiccator. If you are going to use magnetic stirring, do not forget to dry the stirrer bar as it will also be covered in a film of moisture.

It is not advisable to grease the joints in glassware that is to be used for reactions, as it will almost certainly find its way into your reaction product. If it is unavoidable, always use the minimum quantity possible and remove it carefully from the joint immediately after the reaction has been quenched. This can be done using a tissue dampened with chloroform or petroleum ether. It is important to do this before addition of an extraction solvent or before pouring the reaction mixture out through the joint. Rather than using grease, it is much better to use PTFE tape or PTFE sleeves. These are at least as efficient and will not contaminate your product. It is also advisable to seal the outside of the joint by wrapping PTFE tape around to cover the "join." This will help to prevent any condensation on the outside of the glassware from finding its way into the reaction. Joint clips (e.g., Bibby clips) should also be used on all accessible joints to ensure that connections remain sealed.

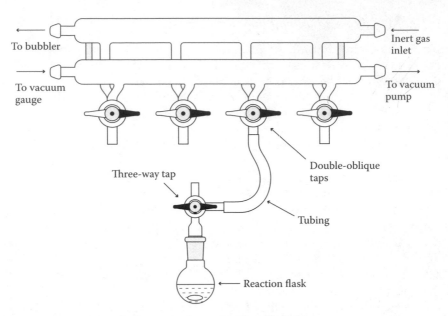

Figure 9.1 Reaction flask attached to a double manifold.

9.2.4   Typical reaction setups using a double manifold

A double manifold (Figure 9.1) is an extremely useful piece of apparatus, especially for carrying out reactions under inert atmosphere. One barrel of the manifold is connected to a high vacuum pump (or house vacuum system) and another to a source of dry inert gas. Its main advantage is that any vessel connected to an outlet can be switched instantaneously from vacuum to inert gas atmosphere or vice versa, simply by turning the two-way tap on the outlets. Several sets of apparatus can be connected at once, the only limit being the number of outlets on the manifold. At the end of the argon (or nitrogen) line, there is a bubbler that prevents air from being sucked back into the system and provides a convenient means of monitoring the gas flow rate. A fast flow is used when filling a system, but once the reaction is set up, one bubble every 5–10 seconds is adequate.

9.2.5   Basic procedure for inert atmosphere reactions

A typical simple reaction setup using a double manifold is shown in Figure 9.1. In this setup, the reaction flask is attached via a three-way tap. The following general procedure shows how this arrangement can be used for a wide range of reactions. Some modifications to this setup are discussed in Sections 9.2.7 and 9.2.8.

Before starting, decide on the order of addition of reagents and if possible arrange them so that any solid reagent is added first. If this is not possible, try to add the solid as a solution as this will be more straightforward than adding a solid. If a reagent really does have to be added as a solid, it can be done using the apparatus shown in Figure 9.9.

Once you have all the apparatus available and have established the best order of addition of the reagents, the reaction can be set up:

1. *Set up and dry the system*
   Connect the dry reaction flask including stirrer bar to the manifold (Figure 9.1), set the two-way tap on the manifold and the three-way tap on the reaction flask (position A, Figure 9.2a) so that the system is evacuated, and heat the glassware with a heat gun (or Bunsen) for a few minutes. Then, with inert gas flowing through the bubbler at a rapid rate, turn the manifold tap slowly so that the hot apparatus fills with inert gas. Keep an eye on the bubbler while you are doing this to ensure that the inert gas flow is high enough to avoid "suck back." Finally, leave the apparatus to cool.
2. *Add initial reactants*
   Solid reactants (or viscous oils) can be weighed directly into the reaction flask at this stage. With the inert gas flow at a reasonably high level, remove the flask from the system and quickly weigh the reagent into it (for reactive solids, see Section 6.4.5). Then, reattach the flask to the three-way tap and switch the manifold tap to evacuate. If there are traces of solvent on the reactant, these can be removed by leaving the flask under vacuum for a while, otherwise

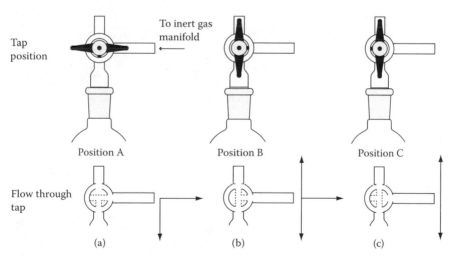

Figure 9.2 Flow through a three-way tap relative to tap position inert gas flows: (a) only into flask, (b) into flask and to atmosphere, and (c) no inert gas flow.

the two-way tap can be switched back to argon straightaway leaving the reagent in the flask under inert atmosphere.

3. *Add solvent*

The best way to add solvent while maintaining the inert atmosphere is to use a syringe (Figure 9.3b). It is best to syringe freshly distilled solvent straight from the collecting head of a still, which is itself under inert atmosphere (Figure 5.2), but anhydrous solvent can also be taken from a bottle (Figure 9.3a) or flask under inert atmosphere. In either case, once you have a syringe loaded with solvent, open the three-way tap to position B (Figure 9.2b), keeping an inert gas flow that is rapid enough to maintain bubbling through the bubbler, and add the solvent. Then, turn the three-way tap back to position A (Figure 9.2a) and reduce inert gas flow to maintain steady bubbling. For large-scale reactions, solvents can be added by disconnecting the flask, adding the solvent quickly and reassembling, but this inevitably leads to air entering the system. With higher boiling solvents, the air can be removed by sequentially partially evacuating and refilling with inert gas several times, using the two-way manifold tap.

It is also possible to degas the solution using the freeze-pump-thaw method. To do this, the solution is placed under a positive

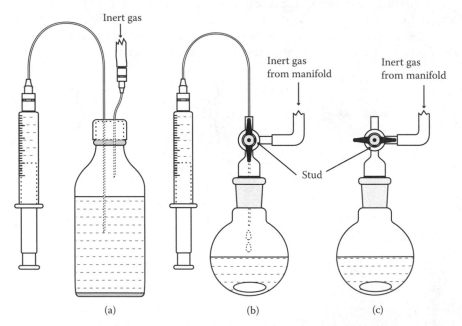

Figure 9.3 Adding air- or moisture-sensitive liquids to a reaction flask: (a) fill syringe, (b) add solvent from syringe, and (c) turn three-way tap to position A (Figure 9.2).

pressure of inert gas and then frozen by immersing the flask in an appropriate cooling bath (Tables 9.3–9.5). While the solvent is completely frozen, the flask is evacuated for 2–3 minutes. The two-way tap on the manifold is then closed and the solvent is then allowed to melt while still under partial vacuum. This freeze-pump-thaw cycle needs to be repeated at least three times to ensure that all the dissolved O_2 has been removed. After the final evacuation, the flask is filled with inert gas before thawing.

4. *Cool the flask*

At this stage, the reaction flask can be immersed in a cooling bath and stirred with a magnetic stirrer. Never cool a flask before it is under the inert atmosphere as this will cause moisture to condense inside the flask.

5. *Add reagents*

It is most commonly the reagents used in the reaction that are air- and/or moisture-sensitive, and consequently, they should be added with great care. After pressurizing the reagent container with inert gas, syringe out the required quantity (Figure 9.3a). Increase the inert gas flow through the manifold so that rapid bubbling is maintained as the three-way tap is opened to position B and then insert the syringe through the tap to deliver the reagent (Figure 9.3b). When all the reagents have been added, turn the three-way tap back to position A and reduce the gas flow so that there is a bubble every few seconds (Figure 9.3c). Any further reagents are added in the same way and the reaction is then left to proceed. To add a cooled reagent, it is advisable to use a cannula made out of PTFE tubing (see Section 6.4 for more details on how to do this).

6. *Reaction monitoring*

Using this setup, it is easy to sample the reaction mixture for monitoring by TLC, HPLC, NMR, and so on. First, increase the inert gas flow through the manifold so that rapid bubbling is maintained as the three-way tap is opened to position B and then insert a long TLC spotter, syringe needle, or Pasteur pipette through the tap to remove a sample of the mixture. Once the sample has been taken, turn the three-way tap back to position A and reduce the gas flow as before.

7. *Work-up*

Once the reaction is complete, it can be worked up as normal (Chapter 10). However, if it is likely that reactive reagents (e.g., organometallics) are still present, be *very careful* when quenching with aqueous solutions. In such instances, it is usually safer to keep the mixture under an inert atmosphere while the aqueous solution is added.

9.2.6   Modifications to basic procedure

The procedure outlined in Section 9.2.5 is best suited for small-scale reactions at or below room temperature where most of the reagents can be added by syringe. Some useful modifications that expand the range of reactions that can be performed are described as follows:

1. *Reactions at elevated temperatures*
   If the reaction needs to be heated, or if there is any possibility that it might be exothermic, a condenser should be incorporated into the reaction setup. Typical setups for reactions that are to be heated are shown in Figure 9.4.

   Note the use of an internal coil condenser in this setup. Most other types of condenser have the cooling water jacket in contact with the outer surface of the condenser. This leads to the possibility of atmospheric moisture condensing on the cold outer surface, then running down onto the ground glass joint, and seeping into the reaction flask. PTFE tape around the joints will help prevent water getting into the reaction vessel, but a coil condenser should prevent the problem all together as all its cooling surfaces are inside the apparatus.

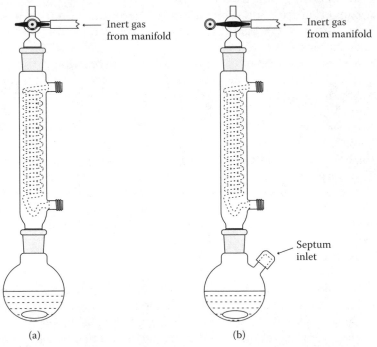

Figure 9.4 Typical setups for inert atmosphere reactions that are to be heated: (a) no septum needed, and (b) septum needed to allow additions via syringe.

The procedure for carrying out the reaction is exactly as described in Section 9.2.5, except that the reaction is heated instead of being cooled. Reagents can be added through the three-way tap, provided a syringe with a long enough needle is used. If this is not possible, a flask with a side arm fitted with a septum (Figure 9.4b) can be used. While the temperature of a reaction is changing, the gas bubbler in the system should be continually checked to make sure there is a constant inert gas pressure. Extra care should be taken to ensure that the rates of reactions at ambient or elevated temperatures are kept under control. Unless you know that the reaction is slow and it is not exothermic at the desired reaction temperature, it is unwise to add reagents quickly, as this could result in an uncontrollable exotherm when the reaction starts. Instead, it is safer to add the rate-limiting reagent at such a rate so as to maintain the rate of reaction and keep the heat generated under control. Using an internal temperature monitoring device is always very useful for gaining a better understanding of thermal attributes of a reaction. Temperature changes can also provide information on reaction rates and indicate when a reaction is complete. These precautions are particularly important when working on a larger scale and more information on this is provided in Chapter 13.

2. *Slow addition of reagents and large-scale work*
When slow addition of an air-sensitive reagent is required, it is best to incorporate a pressure-equalizing addition funnel into the apparatus, and for large-scale work, this is always best. The apparatus can be set up with or without a condenser and can be stirred either magnetically or mechanically. A typical arrangement is shown in Figure 9.5. If very precise slow addition is required, the use of a syringe pump (Figure 4.16) is strongly advised.

Computer control software can be either purchased or written to enable monitoring of reaction parameters such as temperature, pH, stirring rate, and gas evolution. This software can be used to control devices such as syringe pumps, heat–chill systems, and stirrers, based on the measured parameters such as reaction temperature or pH. For example, the rate of addition of a reagent from a syringe pump could be controlled so as to ensure that the temperature of an exothermic reaction does not exceed a set level (see Chapter 13 for more information about reaction scale-up).

To dry this larger, more elaborate setup, it is best to oven-dry the equipment and assemble it hot. It should then be allowed to cool under a stream of inert gas. This can be done by connecting to an inert gas supply via the three-way tap and using a short needle inserted through the septum on the dropping funnel as a vent (Figure 9.6a). Once the apparatus is cool, reactants can be added

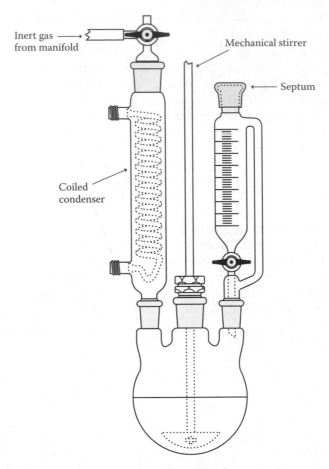

Inert gas from manifold

Mechanical stirrer

Septum

Coiled condenser

Figure 9.5 Larger-scale apparatus for inert atmosphere reactions.

as described in the preceding point. A syringe can be used to add small quantities of reagent to the addition funnel. For larger quantities (>50 cm³), the cannulation technique described in Section 6.4 for bulk transfer of liquids is the best. A graduated dropping funnel is particularly useful in this instance as it allows a known amount of the reagent to be added straight from the reagent bottle (Figure 9.6b). If a mechanical stirrer is used, a PTFE-sealed guide should be used for the stirrer rod. The reaction is then carried out exactly as before.

3. *Addition of a reagent prepared separately or a solution of a solid reagent*
   When a solid reagent is to be added to a reaction setup that is already under inert conditions, it is easiest to add it in solution. To do this, weigh the solid into a dry flask, fit the flask with a three-way tap (Figure 9.2), and connect to a separate outlet on the manifold. Evacuate and fill with inert gas and then with the gas flow high enough to

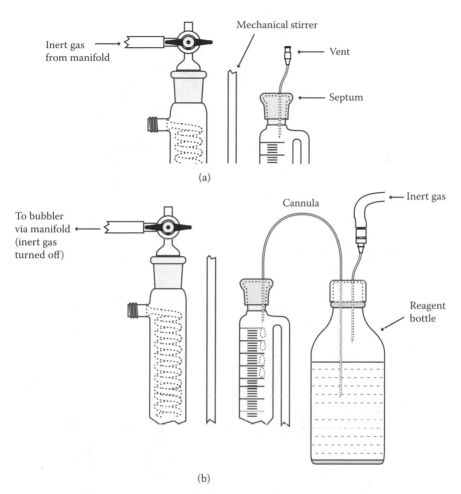

Figure 9.6 Setting up larger-scale apparatus for inert atmosphere reactions. (a) Cooling apparatus under a flow of inert gas. (b) Filling dropping funnel using cannula.

maintain rapid bubbling at the bubbler and open the three-way tap to position B and syringe in the required dry solvent. With the three-way tap turned back to position A, the solution is under inert atmosphere and can be kept like this until required (Figure 9.7). To transfer the reagent, make sure that the inert gas flow is high enough to maintain rapid bubbling while the three-way tap on the reagent vessel is opened to position B. Then, syringe out the solution, turn the tap back to position A, open the tap on the reaction flask to position B, and add the reagent. The same procedure can be used to add a reactive reagent (e.g., Grignard or LDA) that has been prepared in a separate flask.

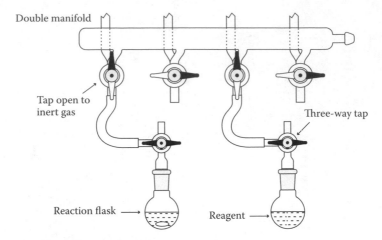

Double manifold

Tap open to
inert gas

Three-way tap

Reaction flask ⟶

Reagent ⟶

Figure 9.7 Using a double manifold.

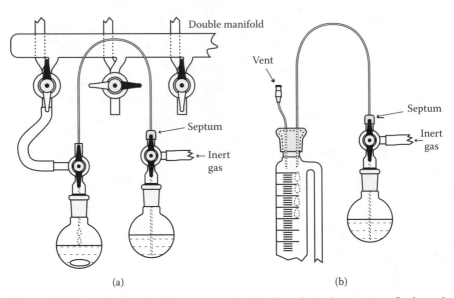

(a)   (b)

Figure 9.8 Transferring liquids via cannula: (a) directly to the reaction flask, and
(b) to an addition funnel.

For large-scale transfers, or for the transfer of a precooled solution, a cannula can be used. To use a cannula, first fit the three-way
tap of the reagent flask with a septum, then pressurize the flask with
a separate inert gas supply, turn the three-way tap to position B,
insert a cannula, and transfer the solution either to the reaction flask
directly (Figure 9.8a) or to an addition funnel (Figure 9.8b).

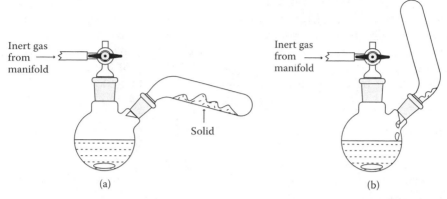

Figure 9.9 Using a solid addition tube: (a) before addition of solid to the reaction, and (b) addition of the solid.

4. *Direct addition of a solid to a reaction under inert atmosphere*
    Although it is much easier to add solids in solution, on some occa-sions, a solid will have to be added directly. When the apparatus is connected to a manifold and with argon as the inert gas, there are several ways of doing this. The simplest procedure involves using a reaction flask with a stoppered side arm. After first increasing the inert gas flow to give very rapid bubbling, the stopper can be removed and the solid reactant added via the side arm. If this pro-cedure is used, great care needs to be taken. The stopper should be removed for the minimum time required to add the reagent, and you should make sure that it is not added at a rate that might lead to the reaction going out of control. Also, if the solid is a fine powder, it is important to make sure that the gas flow does not blow the pow-der out of the flask. Alternatively, an inverted funnel (Figure 6.20) can be used to blanket the apparatus with an inert atmosphere to protect the reaction mixture.
    If the reaction is particularly air sensitive, a better method is to use a solid addition tube. This is a bent tube containing the solid, which is attached to the side arm of the flask (Figure 9.9a) before the system is placed under inert atmosphere. It can then be left in place until the solid needs to be added. At this point, the tube is rotated to allow the solid to pour into the reaction flask (Figure 9.9b).

9.2.7   *Use of balloons for holding an inert atmosphere*

Although we recommend using the double-manifold techniques des-cribed in Section 9.2, reactions can also be kept under inert atmosphere by using a balloon filled with inert gas. This can prove particularly useful if

a double-manifold system is not available. To do this, you first fill the balloon with the inert gas from a cylinder and then attach it to the reaction
system using either a three-way tap or a needle/septum (Figure 9.10),
taking care to ensure that there are no leaks in the system. The balloon
keeps the whole system under a positive pressure of the inert gas and
also allows liquid materials to be added to, or removed from, the flask
via syringe insertion through the septum. If the reaction is to be heated,
a condenser may also need to be included in the reaction setup. If this is
the case, the balloon must be attached at the top of the condenser; otherwise, it will be exposed to hot solvent vapor and will certainly burst. In
this case, it may be necessary to use a two-necked flask so that materials can be added directly to the flask rather than from the top of the
condenser.

With the three-way tap attachment, a vacuum line can be connected
(Figure 9.10a). The system can then be purged by sequentially evacuating and then filling with the inert gas from the balloon. This procedure
ensures that the flask is filled with inert gas, and so, both nitrogen and

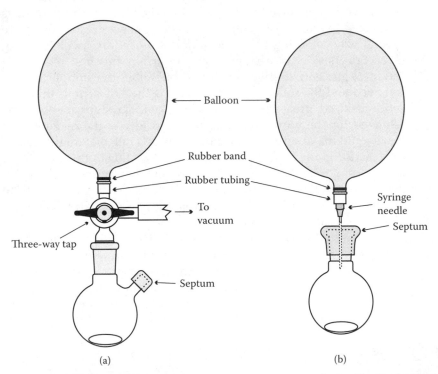

(a)   (b)

Figure 9.10 Using a balloon to maintain an inert atmosphere: (a) attached via
three-way tap to permit evacuation/filling cycles, and (b) attached via a needle
and septum.

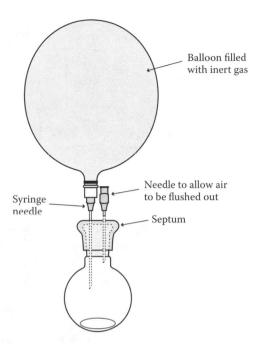

Balloon filled with inert gas

Syringe needle

Needle to allow air to be flushed out

Septum

Figure 9.11 Using a balloon to flush a flask with inert gas.

argon can be used with this setup. Liquid materials can then be added to the flask by syringe via the septum inlet as required.

In the case of the needle/septum attachment (Figure 9.10b), any air contained in the reaction flask can be flushed out by using another needle to vent the system (Figure 9.11). After the reaction flask has been thoroughly flushed with the inert gas, the extra needle is removed and the reaction flask is ready for use. Argon is generally preferred when using this technique because it is more dense than air and so will push the air out more effectively than nitrogen.

Balloons attached to needles are also an extremely useful means of maintaining a positive pressure of inert gas when syringing air-sensitive materials from bottles or containers fitted with septa (Figure 9.3).

Balloon attachments are easily constructed as outlined in Figure 9.12. First, the balloon is pushed over a piece of rubber tubing, which is about the same diameter as the balloon neck. As balloons are perishable, it is advisable to use a "double-balloon system." This is simply two balloons, one inside the other, allowing the inert atmosphere to be maintained even if one bursts. The balloons are secured in place using a piece of wire, elastic band, or Parafilm. The open end of the tubing can then be attached to a three-way tap or to a needle-tubing adapter and needle as shown. It is advisable to seal all joints with PTFE tape or Parafilm to ensure that there are no leaks.

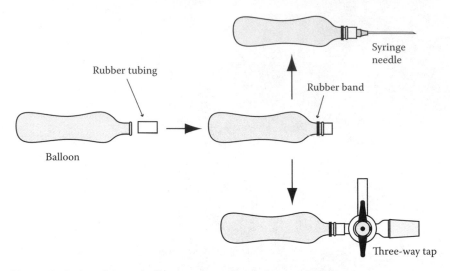

Figure 9.12 Attaching a balloon to a needle or three-way tap.

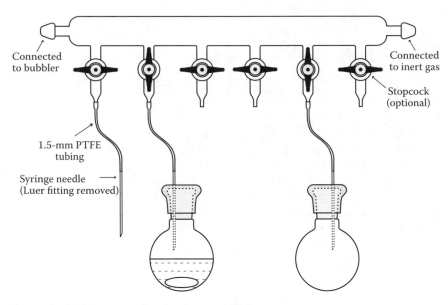

Figure 9.13 Using a spaghetti tube manifold.

9.2.8 Use of a "spaghetti" tubing manifold

The main drawback with balloons is that they have a tendency to perish and burst. A spaghetti tubing manifold (Figure 9.13) will provide a similar low pressure inert gas source, but will be more reliable as long as it is well maintained. It is particularly useful if you need to set up several small-scale reactions running in parallel to one another. This is often the case if you need to optimize the conditions for a reaction.

The reaction is set up in exactly the same way as described in Section 9.2.7, the only difference being the source of the inert gas pressure.

9.3 Reaction monitoring

Most people who are new recruits to the research labs learned their basic skills for carrying out reactions in an undergraduate laboratory. Inevitably, most of the organic chemistry undertaken in these laboratory classes involves following "recipes," which have been well tried and tested. The conditions and the time required for these reactions to reach completion are usually well established, and so, work-ups can be carried out after a preset time. Unfortunately, the assumption that you can predict the time it takes for a reaction to reach completion is a very bad habit to carry over into a research environment.

Even slight changes to a protocol can have a dramatic effect on the time it takes for a reaction to reach completion. Consequently, *every reaction you carry out should be monitored.* If you wish to be an effective researcher, you need to be able to determine exactly when a reaction has reached completion. So, before starting any reaction, it is essential to determine a suitable method for monitoring its progress. Even if you are following a literature procedure, reaction monitoring will usually save you time as well as give you confidence about what is happening. Carrying out a reaction without monitoring its progress is like trying to thread a needle with your eyes closed.

In this section, we have not endeavored to give a comprehensive review of methods that can be used for reaction monitoring, but we will describe the most universal and commonly used modern techniques. Various other monitoring methods can be devised for specific reactions, and sometimes, you may have to use a little ingenuity to find an appropriate monitoring technique for your reaction.

9.3.1 Thin layer chromatography

TLC is a simple but extremely powerful analytical tool. However, it may take a little time before your expertise reaches a consistently high level. This is because a certain amount of intuition is involved in choosing the appropriate solvent system, spotting the correct amount of sample, and so on. Once you have gained experience and confidence in the use of TLC, you will find it extremely useful for a wide variety of purposes.

TLC is often the simplest and quickest way to monitor a reaction. It allows you to follow how the reaction is progressing and to assess the best time to work it up. It can also give a good indication of the purity of a substance. Once you have established a good TLC system for your reaction product, it will also allow you to determine the solvent system and quantity of silica required for flash chromatography (see Figure 11.18). For these reasons, it is always good practice to try and establish a good TLC system for each reaction you do. It is also important that you accurately record the TLC data in your laboratory book (see Chapter 2) so that you, or someone following up your work, can see exactly how the reaction progressed.

There are a wide variety of commercially available TLC plates, but ones coated with either silica or alumina, tend to be the most useful for organic chemistry. Silica plates are the most commonly used; they are slightly acidic, but are suitable for running a broad range of organic molecules. Alumina plates are slightly basic and are often used when a basic compound will not run very well on silica. Most of the information in this section will refer directly to silica plates, but the same principles apply when using alumina plates.

TLC plates typically have either a glass, aluminum, or plastic backing coated with a thin layer of silica, which contains a binding agent to keep it fixed to the backing. Although TLC plates can be homemade, most people prefer to use commercial ones as they give consistent results. For analytical purposes, plates with a 0.25 mm layer of silica are normally used. They are available in a variety of sizes, although 5 × 20 cm (height × width) is probably the most convenient. These can be cut to the required width. A height of 5 cm normally gives adequate resolution, with a very short running time. Plastic- or aluminum-backed plates can easily be cut into strips with scissors, but glass-backed plates seem to give better resolution and can be heated more vigorously for visualization purposes.

To cut a glass TLC plate, place it face down on a clean piece of paper, hold a ruler firmly along the proposed cut, and draw a sharp diamond glass cutter along the line *once only*. Then, holding the plate with the forefinger and thumb of each hand, on either side of the score line, snap the plate along the line. With practice, and a good glass cutter, you should be able to cut plates down to about 1.5 cm width without ever spoiling them.

The following steps outline how best to follow a reaction by TLC:

1. Cut a series of TLC plates to the appropriate size. This would normally be 5 cm high and wide enough such that a 0.5 cm gap can be left between each spot. If you are monitoring a single reaction, the ideal TLC plate size would be 5 × 2 cm.
2. Make a batch of TLC spotters by drawing out Pasteur pipettes to about 0.5 mm internal diameter using a Bunsen. The spotters can be reused many times provided you wash them well with clean solvent in between runs.

3. Prepare a TLC tank by lining with a filter paper and adding your chosen solvent system to a depth of about 0.3 cm. Commercial TLC tanks can be used, but it is usually much cheaper and more convenient to use a 100 cm^3 beaker with a watch glass or Petri dish lid (Figure 9.15b).

4. Make up a TLC sample (ca. 1–2% solution) of your starting material(s) by dissolving a small quantity (<1 mg) in a volatile solvent. A nonpolar solvent is preferable, but dichloromethane is often used because it will dissolve most organic compounds.

5. Using a pencil, gently mark a series of evenly spaced spots along a horizontal line 0.5 cm apart and 0.5 cm from the base of the TLC plate. The number of spots you will need will depend on how many samples you are going to put on the TLC plate. Ideally, this should include all your organic starting materials, co-spots, and reaction mixture(s). The example shown in Figure 9.15 is for a single reaction involving a single starting material. In this situation, three spots are required, and a small amount of the starting material solution should be placed on the first two spots (Figure 9.15a).

6. Next you need to sample the reaction mixture. In many instances, this can be done directly, simply by dipping the TLC spotter in the reaction solution. If the reaction is under an inert atmosphere, you will need to take a TLC sample without allowing air into the apparatus. If you are using the reaction setup shown in Figure 9.1, simply turn the three-way tap to position B and then quickly insert the TLC spotter into the reaction mixture, remove, and return the tap to position A. If you are using a balloon or spaghetti manifold setup, the most straightforward way of doing this is to push a syringe needle through the septum on the reaction flask and insert a TLC spotter through it (Figure 9.14). When using this latter method, it is extremely important that you make sure the needle has a wide enough bore to easily accommodate the TLC spotter. If it is a tight fit, the spotter could jam and break the needle. It is also important to ensure that the positive pressure of inert gas is not sufficient to cause the reaction mixture to spray out of the top of the TLC spotter.

    In some cases, it may be necessary to perform a "mini-work-up" before running a TLC of the reaction mixture. In this case, the TLC sample is best extracted using either a Pasteur pipette or a syringe. A discussion of how best to perform a reaction work-up on small amounts of material can be found in Chapter 12.

7. Once a TLC sample of the reaction mixture has been obtained, it needs to be spotted on the second (co-spot) and third (reaction mixture) spots at the bottom of the TLC plate (Figure 9.15a).

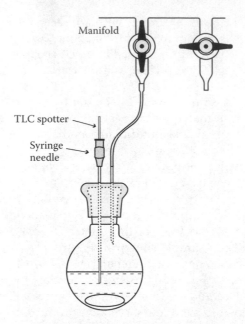

Figure 9.14 Taking a TLC sample from a reaction under inert atmosphere.

8. The TLC plate can then be placed in the TLC tank (Figure 9.15b). Allow the solvent to creep up the TLC plate until it is about 0.3 cm from the top. Then remove it, and mark the level of the solvent front (Figure 9.15c). The distance that an individual compound runs up a TLC plate is extremely variable and depends on the exact conditions under which the plate was run. It is therefore much more informative to run comparative TLCs. This is why when analyzing a reaction mixture, it is essential to include the starting material and a co-spot on the TLC plate. Unless you do this, it will often be very difficult to clearly distinguish between compounds with similar retention times.

Once the TLC has been run, you will need to visualize the spots. There are a number of ways to do this, and in most cases, a combination of techniques will be needed to see all the spots. The three most commonly used techniques are listed below. The first two are nondestructive, and so they should always be carried out in the order shown in the following:

1. The plate can be viewed under a shortwave UV lamp to show any UV-active spots. To do this, you will need to make sure that you use TLC plates that have been pretreated with a fluorescent indicator (normally 254 nm).

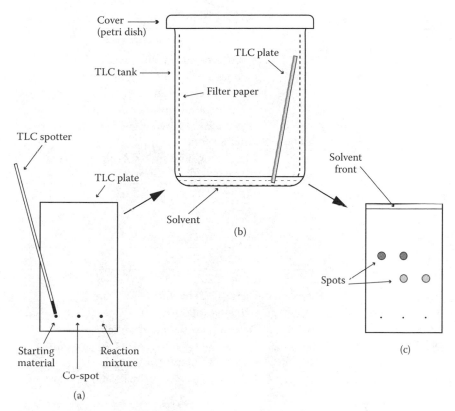

Figure 9.15 Running a TLC: (a) spot the plate, (b) run the TLC, and (c) remove and develop the plate.

2. The plate can be stained with iodine. This can be achieved rapidly by shaking the plate in a bottle containing silica and a few crystals of iodine. The iodine will stain any compound that reacts with it and so is especially good for visualizing unsaturated compounds. Most spots show up within a few seconds, but the stain is not usually permanent.

3. The plate can be treated with one of the reagents listed in Table 9.1 and then heated to stain the spots. The reagents are best stored in small jars that are large enough to allow a TLC plate to be dipped into the solution. To use a stain, first let the TLC solvent evaporate. Then holding the top of the plate with tweezers, immerse the plate as completely as possible in the stain and remove it quickly. Rest the bottom edge of the plate on a paper towel to absorb the excess stain and then heat gently on a hot plate or with a heat gun until the spots appear. This method is irreversible and so should be done last. When glass plates are used, the spots can sometimes be seen more clearly from the glass side of the plate.

Table 9.1 TLC Stains

Stain	Use/comments
Vanillin	Good general reagent, gives a range of colors
Phosphomolybdic acid (PMA)	Good general reagent, gives blue-green spots
Anisaldehyde	Good general reagent, gives a range of colors
Ceric sulfate	Fairly general, gives a range of colors
2,4-Dinitrophenylhydrazine (DNP)[a]	Mainly for aldehydes and ketones, gives orange spots
Permanganate[a]	Mainly for unsaturated compounds and alcohols, gives yellow spots

[a] Does not usually require heating.

Table 9.2 Recipes for TLC Stains

Stain	Recipe
Vanillin	Vanillin (6 g) in ethanol (250 cm^3) + conc. H_2SO_4 (2.5 cm^3)
PMA	Phosphomolybdic acid (12 g) in ethanol (250 cm^3)
Anisaldehyde	Anisaldehyde (6 g) in ethanol (250 cm^3) + conc. H_2SO_4 (2.5 cm^3)
Ceric sulfate	15% aqueous H_2SO_4 saturated with ceric sulfate
DNP	2,4-Dinitrophenylhydrazine (12 g) + conc. H_2SO_4 (60 cm^3) + water (80 cm^3) + ethanol (200 cm^3)
Permanganate	$KMnO_4$ (3 g) + K_2CO_3 (20 g) + 5% aqueous NaOH (5 cm^3) + water (300 cm^3)

The TLC stains that are most commonly used in organic chemistry are shown in Table 9.1. These stains can be made using the recipes shown in Table 9.2.

The distance that a compound travels up a TLC plate is called the retardation factor, which is abbreviated to the R_f value. This is defined by the following equation:

$$R_f = \frac{\text{Distance of center of spot from baseline}}{\text{Distance of solvent front from baseline}}$$

The R_f value of a compound is normally only accurate to about 20%. Consequently, if you need to compare compounds by TLC, it is *essential* to run them on the same TLC plate and to include a co-spot. That said, it is also important to record the R_f value of a compound along with the solvent system used so that anyone following up the work can easily identify an appropriate TLC system.

The R_f value is dependent on the polarity of the compound and the polarity of the solvent used. In general, the more polar the compound, the more tightly it will bind to the silica (or alumina) on the TLC plate. So, less polar compounds will tend to travel further up a TLC plate than more polar compounds in a given solvent. It is often possible to predict whether the product of a reaction will be less or more polar than the starting material. For example, when a ketone or ester is reduced, the resulting alcohol is normally more polar. So, when following a reaction of this type by TLC, the formation of a single lower running spot will usually indicate a successful reaction.

Increasing the polarity of the solvent used for TLC elution will normally cause compounds to move further up the plate. However, the spots will become more diffuse the further they travel up the TLC plate. So, an R_f value of about 0.4 is normally the optimum for analytical purposes. The best TLC solvent system for a particular compound or mixture can only be determined by trial and error. To simplify this process, miscible solvent mixtures comprising a nonpolar solvent and a more polar solvent are generally used. Common TLC solvents can be grouped into three categories based on polarity:

1. *Very polar solvents:* methanol > ethanol > *iso*-propanol.
2. *Moderately polar solvents:* acetonitrile > ethyl acetate > chloroform > dichloromethane > diethyl ether > toluene.
3. *Nonpolar solvents:* cyclohexane, petroleum-ether, hexane, pentane.

The most commonly used solvent mixture for small organic molecules is petroleum ether–ethyl acetate. The polarity of this mixture is easily adjusted by changing the proportions of the two solvents. If the compounds being analyzed are very polar and will not travel in 100% ethyl acetate, then more polar solvents will need to be used. Dichloromethane–methanol is commonly used in this instance. Amines and carboxylic acids are commonly encountered organic compounds that are highly polar. These substances also tend to give streaks rather than distinct spots on silica TLC plates. Including a small amount of acetic acid (1–5%) in the TLC solvent mixture will often help for carboxylic acids, and a small amount of triethylamine (1–5%) will usually help for amines.

The degree of separation between different compounds in a mixture will also vary according to the TLC solvent used. So, if compounds do not separate, or give poor separation, different solvent mixtures should be tried. Better separation of spots can also sometimes be achieved by eluting the TLC plate several times (multiple elutions). This is done by eluting the TLC plate as normal, removing it from the TLC tank and allowing the solvent to evaporate, and then re-eluting the plate as before. Eluting a plate *n* times is effectively the same as running a plate *n* times the length. Occasionally, it may be necessary to run a parallel series of TLCs in different solvent mixtures to resolve all the components of a reaction mixture.

It is also important to be aware that some compounds can decompose in TLC plates. For example, the silica used in silica TLC plates is acidic in

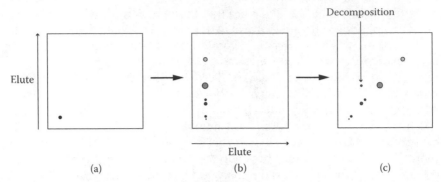

Figure 9.16 Running a two-dimensional TLC: (a) place spot in one corner, (b) elute plate, and (c) turn plate by 180° and elute again.

nature and so using such plates can lead to decomposition (on the plate) of acid-sensitive compounds. It is sometimes possible to get around this problem by switching to alumina TLC plates (these suffer from the disadvantage that resolution is generally not as good, and the plates are basic in nature) or by adding a small amount of an amine (usually ammonia or triethylamine) to the TLC solvent mixture to neutralize the acidic sites on the silica.

If you suspect that a compound may be decomposing on the TLC plate, you can check by running a two-dimensional plate. This is done by cutting a square plate (approximately 5 × 5 cm) and spotting the compound in the bottom left-hand corner (0.5 cm from the bottom) as shown in Figure 9.16a. The plate is then eluted in the usual way to give the spots in a line up the left-hand side of the plate. The plate is then removed from the TLC tank and the solvent allowed to evaporate. It is then placed back in the tank, this time with the line of spots along the bottom, and re-eluted (Figure 9.16b). If this gives the same number of spots now aligned along the diagonal, no decomposition is occurring (or the decomposition is so fast that it was complete before the first run). If decomposition is occurring when the TLC runs, the spot due to the unstable compound will show decomposition products off-diagonal as shown in Figure 9.16c.

It is a useful precaution to carry out this test before any form of preparative chromatography, especially if you suspect that a compound may be unstable to silica.

9.3.2 *High performance liquid chromatography*

There are a broad range of HPLC techniques and many different types of equipment that can be used. It is beyond the scope of this book to describe in great detail the methods for operating the equipment. This section will therefore focus on some of the ways that analytical HPLC can be used to aid the synthetic organic chemist.

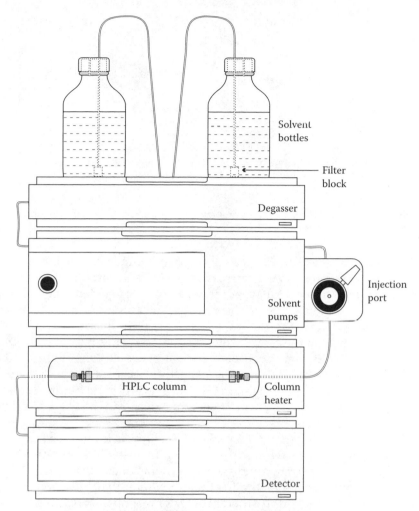

Figure 9.17 A typical analytical HPLC setup.

A typical analytical HPLC system is shown in Figure 9.17. Solvents are pumped from the reservoirs through a degasser to a piston pump that controls the flow rate. From the pump, the solvent passes through a pulse damper that removes some of the pulsing effect generated in the pump and also acts as a pressure regulator.

In between the pulse damper and the column, there is an injection port that allows the sample to be introduced into the solvent stream. In the "load" mode, the solvent bypasses the sample loop, into which the sample is injected from a syringe (Figure 9.18a). This enables the solvent to flow through the entire system before the sample is injected. While the

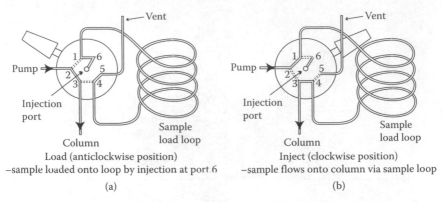

Figure 9.18 Schematic of the injection port in load (a) and inject (b) positions.

injection port is set to "load," a solution of the sample can be introduced into the sample loop using a microsyringe. A 20-μL sample loop is typically used for analytical HPLC. As the sample is injected, it displaces the solvent in the sample loop into the vent. On switching to "inject," the solvent stream is diverted through the load loop, introducing an accurately measured volume of the sample solution onto the column (Figure 9.18b).

Components are separated on a normal phase HPLC column in the same way as they would be on a TLC plate, the less polar compounds running faster and coming through first. The effluent from the column passes through a detector (usually an UV spectrometer), which registers a data peak when a component is present.

The time at which the compound comes off the column is characteristic of that particular material and is referred to as the retention time (R_t). The area under any compound peak is proportional to the quantity of that component and the method is therefore quantitative. However, different compounds will have different detector responses. So, to accurately measure the quantities of reaction components by HPLC, it is essential that you first calibrate the system using samples of all the compounds involved. Since the peak area is proportional to the amount of a compound, it is possible to take a known amount of the compound, make a standard solution, and inject specific quantities to work out the proportionality constant. However, a more accurate method of calibration is to use an internal standard. To do this, follow this procedure:

1. Choose a standard that is a readily available stable compound, with a retention time away from the peaks of interest.
2. Make up at least three mixtures containing varying quantities of the standard (std) and each of the compounds that are to be analyzed.
3. Run the mixtures and measure the areas of each peak.

The mass (M) of material under any peak, y, is

$$M_y = k_y \times \text{Area}_y$$

So, comparing the area under the standard peak with that under the can be represented as follows:

$$M_y / M_{std} = k_y / k_{std} \times \text{Area}_y / \text{Area}_{std}$$

Using this equation, we can work out k_y / k_{std}, which is a constant, known as the correction factor for the compound that is responsible for peak y. Using data from each of the runs, the average correction factor for each compound can be calculated.

4. Now if we want to calculate the quantity of a compound in a mixture, the mixture is spiked with a known quantity of the standard and the following equation is used:

$$M_y = (k_y / k_{std}) / M_{std} \times \text{Area}_y / \text{Area}_{std}$$

For good quantitative results from analytical HPLC (or GC, Section 9.3.3), you should aim to produce chromatographs with symmetrical peaks. Tailing of the peaks is usually caused by overloading and can thus be avoided by reducing the quantity of sample applied. If this does not solve the problem and the tail of a component is long and drawn out, there may be an incompatibility between the compound and the stationary phase, a problem that is less easy to rectify.

Setting up an HPLC system takes a significant amount of time and effort compared with qualitative methods such as TLC. Nevertheless, there are many occasions when it is well worth doing. One reason to use HPLC is that the compounds in which you are interested cannot be separated by TLC (e.g., enantiomers). Another common reason is that you require a quantitative technique. This is often the case when you are trying to optimize a reaction to maximize the quantity of one product over another.

Another important use of analytical HPLC is identification of a compound by comparison with a known substance. Under a specific set of conditions (temperature, solvent, flow rate, and quantity applied), any compound will have a specific retention time on a given HPLC column. Consequently, this can be used as a characteristic of that compound. However, just as a mixed spot should always be run when comparing substances on TLC, so with HPLC, a single enhanced peak should be observed when the comparison substance and the unknown are injected as a mixture. Even then, caution should be used, since a single peak is not absolute proof that compounds are the same.

Preparative HPLC is also widely used in organic chemistry for separating compounds that cannot be easily purified by column chromatography (see Section 11.9 for more details). This is done in the same way as analytical HPLC, but the HPLC columns and the HPLC pump volumes required are much larger. Before committing all your material to a preparative column, it is always best to run a small quantity of the sample on an analytical column to work out the best conditions. Commercially available "matched" analytical and preparative HPLC columns are particularly useful for this purpose as they are directly comparable with one another.

If you are monitoring a reaction by HPLC and need to know the identity of one or more of the products, it is usually possible to isolate 1–5 mg of compound over a few runs on an analytical column. This should be sufficient to get a full range of spectral data on the compound. On simple HPLC systems, this can be done manually by collecting the effluent from the column when the peak of interest is coming off and repeating several times. More sophisticated systems may have an automated sampler and a fraction collector and can be set up to automatically collect a particular peak or peaks over a large number of runs.

In recent years, developments in both analytical LC systems and MS have led to HPLC-MS and UPLC-MS systems that can be used to provide chromatographs together with molecular mass data on the separated components within a few minutes. The molecular mass data provided is particularly valuable and is often sufficient to identify the compound concerned. Simple low-resolution LC-MS systems are now becoming commonplace for reaction monitoring and newer more sophisticated MS detectors (e.g., time-of-flight and MS-MS) are also now being used to provide more detailed structural information including accurate mass measurements.

9.3.3   Gas–liquid chromatography (GC, GLC, VPC)

Gas–liquid chromatography (GC) is also commonly used for reaction monitoring and for analysis of reaction products. It can be used for the analysis of any compounds that are volatile below about 300°C and thermally stable. It is not the intention of this section to give a detailed description of GC instrumentation, but simply to outline some of the uses of the technique for reaction monitoring and related work.

GC is a very sensitive technique requiring only very small amounts of sample (10^{-6} g). A solution of about 1% is sufficient and a few microliters of this are injected into a heated injection port. A stream of carrier gas, usually helium, passes through the injector and sweeps the vapors produced on the column, which is contained in an oven (Figure 9.19). The temperature of the oven can be accurately controlled and can be either kept constant or increased at a specified rate. Separation of the components in GC

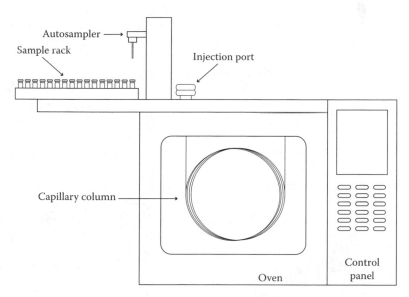

Figure 9.19 A typical GC setup.

is not based on the principle of adsorption, as it is in LC, but on partition. A GC column is rather like an extremely effective distillation column
with the relative volatility of the components being the main factor that
determines how quickly they travel through the column. The stationary
phase of the column is typically a very high molecular weight, nonvolatile
oil, which has a very large surface area, and the gaseous components of
the mixture are partitioned between the oil and the carrier gas at different
rates. Thus, the components are separated along the length of the column
and emerge as discrete bands. The gas stream passing out of the column
enters a FID that produces an electric current when a compound is burned
in the flame. The electric current is amplified to produce a peak, which is
recorded. These detectors are very sensitive and the response produced is
proportional to the quantity of material being burned, thus the peak area
is proportional to the quantity of sample. As with HPLC, the time taken
for a particular substance to reach the detector is characteristic of that
substance and referred to as the retention time (R_t).

Capillary columns are normally used for GC. These are made from
fused silica capillary with an inside diameter of between 0.2 mm and
0.5 mm and are polymer coated. They have no packing, but instead, the
liquid stationary phase is bonded to the inside wall of the capillary, and
this allows gas to flow very easily. Because of this, the columns can be very
long (between 12 m and 100 m) making them very effective for separating most compounds. Capillary columns give extremely high sensitivity
and only a very small quantity of material is required. For this reason, the

injector normally incorporates a "splitter," so that only a small portion of the sample injected actually enters the column. There are a wide variety of stationary phases available ranging from Apiezon greases, which are very nonpolar compared to polyethylene glycols that are very polar. However, the type of stationary phase is often not critical for GC as most capillary columns will give efficient separation. This means that GC instruments can be operated successfully with minimal prior expertise. For most purposes relating to preparative organic chemistry, it is sufficient to rely on just two types of column, one nonpolar (such as a BP1) and another polar column (such as a BP20).

Capillary GC instruments are so simple to use that, provided there is one close by, monitoring a reaction by GC is almost as quick as running a TLC. It is common to turn to GC monitoring when TLC does not provide resolution between starting material and product or between one product and another. GC will usually separate components that co-run on TLC.

GC also allows quantitative analysis and is widely used for determination of product ratios. This makes it an ideal technique for optimization studies, where a large number of small-scale reactions are carried out under different conditions and product ratios are measured simply by syringing out a few microliters from each and then injecting them into the GC instrument. For quantitative studies, the GC instrument can be calibrated in exactly the same way as described for HPLC (Section 9.3.2).

The identity of a compound can also be determined by GC, if an unknown has the same retention time and co-runs with a known compound when the two are injected as a mixture, but just as with TLC and HPLC, caution should be exercised when using this approach.

GC combined with MS (GC-MS) is now routinely used as a very powerful reaction analysis technique. A wide range of bench-top GC-MS machines are commercially available. The mass spectrometer simply acts as the detector, but as well as providing a normal GC trace as a series of peaks a mass spectrum of each individual component is obtained.

An example of how GC-MS can be useful is illustrated by the reaction shown in Figure 9.20. The first time this reaction was carried out, TLC analysis indicated that the starting material had been completely transformed into a single product, which ran as one spot in a range of solvent systems. However, the reaction product did not appear to be pure by ^{1}H NMR spectroscopy. When a GC-MS was run on the reaction product, two compounds (which ran together on TLC) were separated, and from analysis of the mass

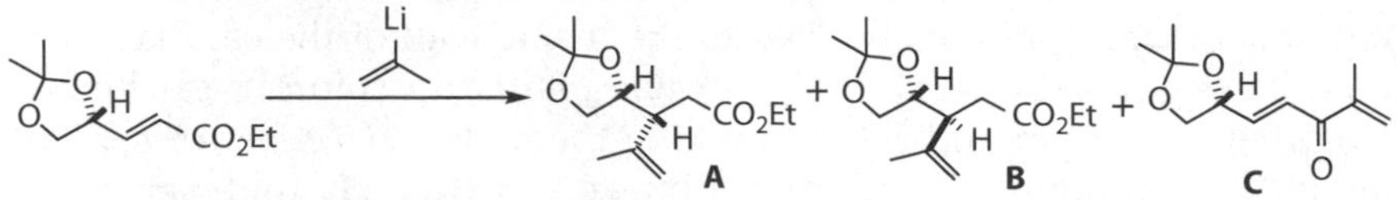

Figure 9.20 Organolithium addition.

spectra, these were identified as compounds A and C. In addition, 3% of an isomeric compound B was also detected and identified using this technique. This minor by-product was seen on TLC due to lack of sensitivity of this technique. Having determined the identity of these reaction products, GC was then used to monitor a series of optimization reactions leading to an improved procedure for generation of compound A.

9.3.4 NMR

NMR is another widely used means of monitoring reaction progress. ^{1}H NMR is particularly useful for monitoring organic transformations because the spectra can be acquired quickly (few minutes) and the characteristic ^{1}H NMR chemical shift and multiplicity of protons within a given structure usually allows easy identification of reaction components. Integration of the ^{1}H NMR signals also allows quantitative analysis of reaction mixtures, although as with HPLC and GC, it is important to use an internal standard and to calibrate your spectra to obtain accurate results.

There are two basic approaches to monitoring reactions by NMR. You can set up a reaction in the normal way and then take samples of the reaction mixture at set time intervals and run NMR spectra of each sample. This can be very effective, but it is important to take into account the following issues:

1. If the reaction mixture is heterogeneous, it will be difficult to ensure that you have taken a representative sample of the mixture.
2. If the sample taken contains all the reaction ingredients, the reaction will still be able to continue. Consequently, your NMR data will relate to the time that the spectrum was recorded, not the time the sample was taken from the reaction mixture. If the reaction is performed below room temperature, it is essential that you "quench" the sample immediately after it is taken, to prevent further reaction.

Alternatively, you can perform the entire reaction inside a NMR tube (see Section 12.4). This has the advantage that you do not need to disturb the system to obtain a spectrum. A wide range of NMR solvents are commercially available, so as long as the reaction mixture is homogeneous, this can be an extremely versatile method of monitoring its progress.

9.4 Reactions at other than room temperature

In many cases, it is necessary to carry out reactions at nonambient temperatures. Usually, reactions that are exothermic, or involve thermally unstable intermediates, have to be carried out at low temperatures, typically in the range 0°C–100°C. Similarly, reactions that are endothermic or have high energies of activation have to be carried out at higher temperatures, typically in the range 30°C–180°C although in some cases temperatures exceeding 300°C may be necessary. This section deals with the techniques involved in such situations.

9.4.1 Low-temperature reactions

In general, all low-temperature reactions should be done under inert atmosphere (nitrogen or argon) to avoid atmospheric moisture being condensed into the reaction mixture. A simple procedure for cooling reactions in round-bottom flasks and similar vessels is to place the reaction vessel in a cooling bath (Figure 9.21). This is done by placing a cooling mixture into a lagged bath and then immersing the reaction vessel in the cooling mixture to a depth that ensures the reaction contents are below the level of coolant. The temperature of the coolant can be monitored by means of a low-temperature (alcohol) thermometer immersed in the bath. Better temperature control over prolonged periods can be achieved by using a jacketed vessel or a Peltier controlled device, as described in Section 4.6. See also Chapter 13, where the use of jacketed vessels for larger-scale work is described.

It is important to note that when you perform a reaction in a cooling bath, the reaction mixture will not necessarily be at the same temperature as the coolant. Poor thermal transfer and exothermic processes will likely result in the reaction mixture being at a higher temperature than the cooling bath. So, where possible, the internal reaction temperature should also be monitored. A particularly convenient way of doing this is to use a

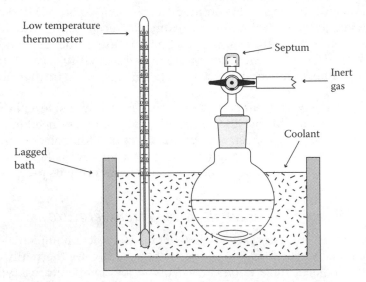

Figure 9.21 Using a cooling bath.

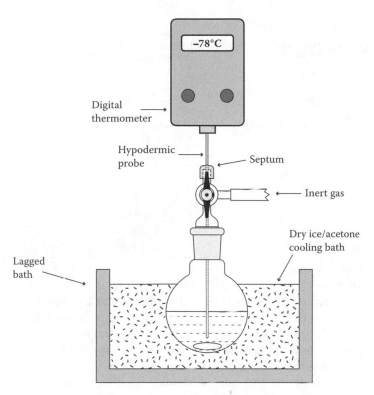

Figure 9.22 Monitoring internal temperature using a digital thermometer.

digital low-temperature thermometer. These are commercially available, and most come with a hypodermic probe, which can be inserted into the reaction flask through a septum (Figure 9.22). This is often a much easier procedure than the alternative method of setting up the reaction apparatus to include an internal thermometer and is especially useful for small-scale setups where an internal thermometer cannot be used. The output from some digital thermometers can be captured by a computer to provide a temperature profile over the course of the reaction.

The three main types of cooling mixtures are described as follows:

1. *Ice–salt baths.* Various salts or solvents can be mixed with crushed water ice to produce subzero temperatures. In practice, temperatures ranging from 0°C to −40°C can be obtained (Table 9.3). However, at temperatures below −10°C, these cooling mixtures consist of granular ice–salt particles with little or no liquid. This leads to poor thermal contact with the reaction vessel, so if careful temperature control is important, it is better to use a liquid or slush coolant in these instances.

Table 9.3 Ice-Based Cold Baths[1]

Additive	Ratio (ice/additive)	Temperature (°C)
Water	1:1	0
NaCl	3:1	−8
Acetone	1:1	−10
$CaCl_2.6H_2O$	4:5	−40

Table 9.4 Dry Ice Cold Baths[2]

Solvent	Temperature (°C)	Solvent	Temperature (°C)
Ethylene glycol	−15	*m*-Xylene	−47
o-Xylene	−29	Chloroform	−61
Heptan-3-one	−38	Ethanol	−72
Acetonitrile	−42	Acetone	−78

2. *Dry ice–solvent baths.* Solid carbon dioxide (dry ice or cardice) is commercially available as pellets or blocks and forms effective good cooling mixtures when combined with an organic solvent (Table 9.4). The cooling baths are prepared by adding the dry ice pellets carefully to a bath containing the requisite solvent until the temperature required is reached.

The temperatures quoted in Table 9.4 refer to baths in which an excess of dry ice has been added to the solvent. In this case, cooling mixtures ranging from −15°C to −78°C can be achieved.

3. *Liquid nitrogen slush baths.* Slush baths are made by adding liquid nitrogen carefully to a solvent contained in the bath, with continuous stirring (use a glass rod or metal spatula, not a thermometer!). The coolant should become the consistency of ice cream, and stirring should prevent any solidification. Again, a variety of liquids can be used to give temperatures ranging from 13°C to −196°C (Table 9.5). Such cooling systems can often be left for several hours if the cooling bath is well lagged; however, for longer periods (overnight), some form of mechanical cooling is usually necessary. In such instances, the reaction vessel can be placed in a refrigerator or cooled by the use of a portable commercial refrigeration unit.

9.4.2   Reactions above room temperature

Reactions above room temperature usually require modifications to the standard equipment setup. In some instances, the reaction can be performed in a sealed tube (Carius tube), usually made of thick-walled glass. The reaction mixture is placed in the tube, which is then sealed (Figure 9.23), placed in an oven, and heated to the appropriate temperature. After the

Table 9.5 Liquid Nitrogen Slush Baths[3]

Solvent	Temperature (°C)	Solvent	Temperature (°C)
p-xylene	13	Chloroform	−63
p-dioxane	12	Isopropyl acetate	−73
Cyclohexane	6	Butyl acetate	−77
Formamide	2	Ethyl acetate	−84
Aniline	−6	2–butanone	−86
Diethylene glycol	−10	*Iso*–propanol	−89
Cycloheptane	−12	n–propyl acetate	−92
Benzyl alcohol	−15	Hexane	−94
o-dichlorobenzene	−18	Toluene	−95
o-xylene	−29	Methanol	−98
m-toluidine	−32	Cyclohexene	−104
Thiophene	−38	*Iso*–octane	−107
Acetonitrile	−41	Ethanol	−116
Chlorobenzene	−45	Methylcyclohexane	−126
m-xylene	−47	n–pentane	−131
Benzyl acetate	−52	*Iso*–pentane	−160
n-octane	−56	Liquid nitrogen	−196

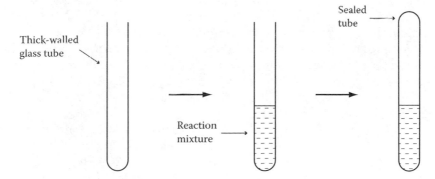

Figure 9.23 A simple sealed tube (Carius tube).

reaction is complete, the tube is cooled, opened, and the contents removed. Such a technique is employed when temperatures greater than the boiling point of solvent are required or for reactions involving extremely volatile compounds. This technique is hazardous and requires a high degree of skill, since heating leads to a pressure buildup inside the tube, which can result in an explosion if there are any flaws in the seal.

An easier-to-use alternative to the all-glass-sealed tube is a reaction tube. This is a thick-walled glass tube with a PTFE screw seal at the top

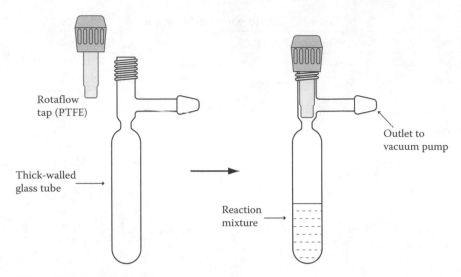

Rotaflow
tap (PTFE)

Outlet to
vacuum pump

Thick-walled
glass tube

Reaction
mixture

Figure 9.24 A reaction tube.

(Figure 9.24). It has the advantage of being reusable, and commercial versions of this apparatus are available. It also incorporates a side arm that allows evacuation or purging with an inert gas before sealing the tube. It should be noted that some apparatus of this type use an O-ring seal. In such instances, it is essential to make sure that the O-ring is made of a material inert toward the reaction contents, otherwise the seal may fail.

For most reactions above room temperature, an open system that does not lead to a buildup of pressure is employed. This usually consists of a reaction vessel protected with a condenser (Figure 9.25). The condenser is used to prevent the evaporation of volatile materials (usually the solvent) from the reaction mixture.

There are many different designs of condenser available, and the type used depends on the nature of the reaction involved. The most common designs of condenser are the coil condenser (Figure 9.26a), the Liebig condenser (Figure 9.26b), the double-jacketed coil condenser (Figure 9.26c), and the cold-finger condenser (Figure 9.26d). Other condensers available tend to be simple modifications of these four types. The Liebig condenser, the coil condenser, and the double-jacketed condenser are similar in design and function. They are water-cooled via connection to a cold water tap; in the case of the Liebig condenser, the water flows in at the bottom and flows out at the top giving a jacket of cold water around the condenser stem and leading to a cold surface on the inside. Any volatile materials in the reaction condense on the cold outer surface and run back into the reaction mixture. The coil condenser functions in a similar way except that the cold surface is now on the inside of the condenser. This can offer an advantage in humid

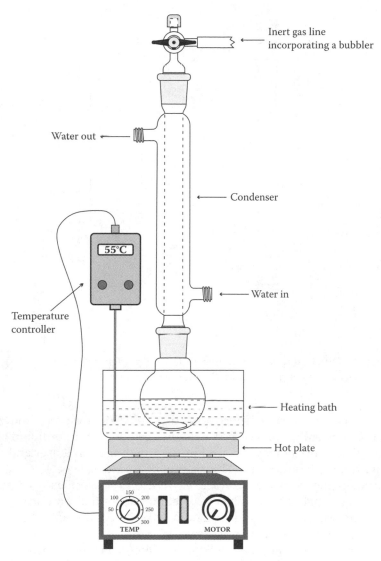

Figure 9.25 A typical setup for performing a reaction at reflux.

locations because there is less tendency for atmospheric moisture to condense on the outside of the condenser and run down over the reaction vessel.

The double-jacketed coil condenser is also water-cooled; again water flows in at the bottom and out at the top. This condenser design tends to be more efficient than the other two because it provides a greater area of cold surface. Consequently, it is preferred when low boiling materials (≤40°C) are involved.

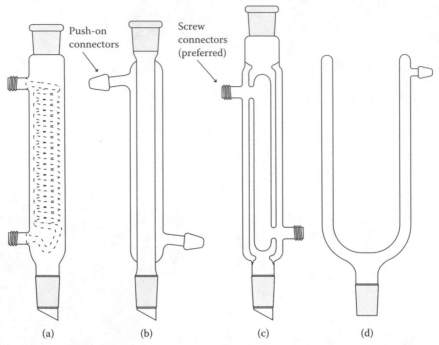

Figure 9.26 Different types of condensers: (a) coil condenser, (b) Liebig condenser, (c) double-jacketed condenser, and (d) cold-finger condenser.

The cold-finger condenser is rather different from the preceding three. It is cooled by either solid carbon dioxide/acetone (−78°C) or liquid nitrogen (−196°C). The coolant is placed in the top of the condenser and more coolant is added as required. This results in an extremely cold surface on the inside of the condenser. Condensers of this type are usually employed for reactions that involve solvents or components that boil at or below room temperature (e.g., liquid ammonia, b.p. −33°C), although they can be used for higher boiling materials as well.

As with low-temperature reactions, a common method of increasing the temperature of a reaction is to place the reaction flask in a bath (Figure 9.25). In this case, the contents of the bath are then heated, usually using a stirrer-hot plate fitted with a temperature controller. There are five commonly employed heating baths: water, silicone oil, Wood's metal, flaked graphite, and sand. Water baths typically consist of a Pyrex glass container filled with water and are used for reactions requiring temperatures up to 100°C. Silicone oil baths are similar in design but can be used for temperatures up to about 180°C, although the maximum temperature in this case depends on the precise type of oil employed. Wood's metal is a commercially available alloy (50% Bi, 25% Pb, 12.5% Sn, and 12.5% Cd) that melts at 70°C. It has excellent thermal properties and so is a safe material for use at quite high temperatures (up to 300°C). It is normally used in a steel container and heated using a hot plate. Similarly, flaked graphite and sand can be used in heating baths up to 300°C, although they cool slowly compared with Wood's metal.

There are also a wide range of aluminum heating blocks that are now commercially available. These come in a wide variety of shapes and sizes and usually have a hole so that a temperature probe can be inserted into the block (Figure 9.27). When using aluminum heating blocks, it is essential

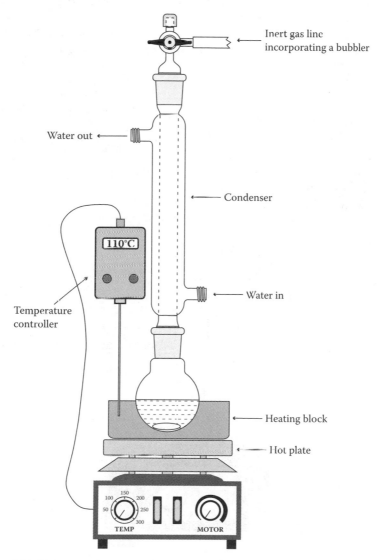

Figure 9.27 Using an aluminum heating block.

to make sure that you have the correct size glassware, otherwise the thermal contact between the block and the flask will be inefficient. Aluminum heating blocks are normally used for temperatures up to 250°C.

When using heating baths or heating blocks in conjunction with stirrer-hot plates, it is essential to attach a temperature-controller probe to the stirrer-hot plate to ensure the heating is switched off once the desired temperature has been reached.

Reactions can also be heated by an electric heating mantle (Figure 9.28), although this is generally less satisfactory since it is more difficult to control the temperature of the mantle surface and excessive heating of the reaction flask can result. Because of this, heating mantles should be avoided wherever possible and should not be used for reactions below 100°C.

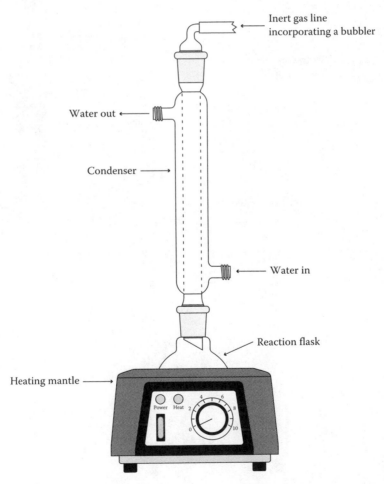

Figure 9.28 Using a heating mantle.

Just as with low-temperature reactions, better temperature control over prolonged periods can be achieved by using a jacketed vessel or a Peltier controlled device, as described in Section 4.6. See also Chapter 13, where the use of jacketed vessels for larger-scale work is described.

9.5 Driving equilibria

A number of important organic reactions involve equilibria and do not give good yields unless the equilibrium can be shifted in favor of the product. Equilibrium can be driven toward the product(s) by using an excess of one of the reactants, by continuously removing one of the products, or by changing the temperature or the pressure at which the reaction is carried out. In most cases, use of excess reagent or removal of a product can be achieved using normal apparatus and techniques. This section outlines two additional types of apparatus that are commonly used in this context: Dean–Stark traps and high pressure reactors.

9.5.1 Dean–Stark traps

Perhaps the most commonly encountered equilibrium reactions are those involving water as a reactant or product. Driving such equilibria by using excess water (e.g., hydrolysis reactions) is easy, but driving equilibria by removing water (e.g., in ester or acetal formation) can be more difficult. An excellent device for the continuous removal of water from a reaction mixture is the Dean–Stark trap.

The apparatus is assembled as shown in Figure 9.29 and the reaction is conducted in a solvent that forms an azeotrope with water (usually a hydrocarbon such as toluene). When the mixture is heated, the solvent/water azeotrope distills and on condensing is collected in the trap. The water then separates and the less dense organic solvent flows back into the reaction flask. It is usually easy to monitor the progress of the reaction either by recording the volume of water produced or by waiting until the characteristically milky heterogeneous azeotrope is no longer produced. A Dean–Stark trap can also be used to remove volatile alcohols, such as methanol and ethanol, even though these are miscible with most organic solvents. In this case, the alcohol can be removed by absorption into 5A molecular sieves placed in the trap. The same effect can also be achieved using a Soxhlet extractor (Figure 10.2) containing molecular sieves.

On a very small scale, simply placing some activated 4A molecular sieves in the reaction flask can be an efficient means of removing water. This method is effective in driving equilibria and is also used to protect reactions that are adversely affected by water.

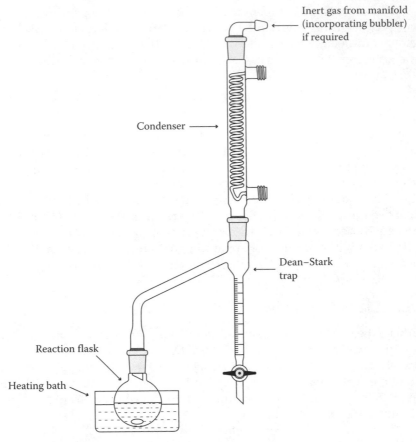

Figure 9.29 Using a Dean–Stark trap.

9.5.2   *High-pressure reactions*

A more esoteric technique is to carry out the reaction at very high pressures (>10 kbar), thus shifting the equilibrium to the side of the components that have the smaller volume. Some transformation of the products must be carried out as soon as they are removed from the reactor, to prevent re-equilibration. More details on the applications of high-pressure techniques are available in literature.[4]

9.6   *Agitation*

Most reactions require some form of agitation to ensure efficient mixing. There are numerous reasons why agitation is importance, including to facilitate mass transfer in heterogeneous and/or multiphase mixtures; to provide efficient mixing of a reagent/reactant during addition; and to allow heat to be dissipated through the reaction medium. Appropriate agitation can be crucial in ensuring that the desired reaction occurs at a reasonable rate and/or with good selectivity. If changing the scale of a reaction (to either larger or smaller scale) leads to a changed outcome or rate, this is often an indication of ineffective agitation. It is therefore important that you employ an agitation technique that is appropriate for the reaction and the size of the reaction. The most commonly used methods of agitation are outlined in this section.

9.6.1   Magnetic stirring

Magnetic stirrer machines (Figure 9.30) are commonly available and come in two general types: either a simple stirring machine or one that also incorporates a hot plate. They consist of a box containing a motor that drives a magnet, which spins horizontally. Most magnetic stirrers have a flat top to allow cooling or heating baths to be placed on top. Agitation of the reaction mixture is achieved by placing a magnetic stirrer bar (also called a magnetic follower or flea) in the reaction mixture. The reaction vessel is then clamped over the top of the stirrer machine in such a position so as to allow the mixture to be stirred by magnetic interaction of the follower with the magnet in the stirrer machine. Usually, the rate of stirring can be adjusted between 0 and 4000 rpm using a control on the stirrer machine.

There are also variations on this design that allow multiple flasks or reaction vials to be stirred using a single magnetic stirrer machine. More sophisticated versions also incorporate heating or cooling blocks to control the temperature of all the reactions. These can be very useful, especially if fume hood space is an issue.

Magnetic followers consist of a magnet coated with an inert polymer, usually PTFE or PVC, and come in a variety of shapes and sizes

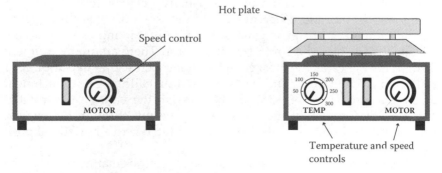

Figure 9.30 Magnetic stirrer machines.

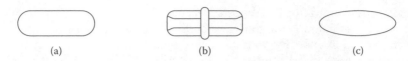

Figure 9.31 Magnetic followers: (a) bar, (b) octagonal, and (c) egg-shaped.

(Figure 9.31). It is important that the polymer coating does not react with any components in the reaction mixture, and because of this, PTFE is the recommended coating for most reaction systems. One notable exception to this guideline is reactions involving metal–ammonia solutions that attack PTFE. The size of the follower is also important; it should be large enough to stir the reaction mixture effectively but not so large that it will not sit flat in the bottom of the reaction flask.

Because the follower is driven by a magnetic field and has no mechanical connection with the stirrer machine, this method of agitation does not require special modification of a reaction apparatus.

Magnetic stirrer machines are probably the most commonly employed method of agitation for organic reactions, but can become ineffective if particularly viscous systems are encountered. They may also be ineffective if the reaction vessel has to be placed inside another piece of apparatus such as a heating mantle or a large cooling bath. In such cases, the extra apparatus can effectively shield the reaction flask from the magnetic field created by the stirrer machine. Magnetic stirring can also be a problem for large-scale reactions (reaction volumes over 1 L), and in such cases, mechanical stirring is usually preferred.

9.6.2 *Mechanical stirrers*

Mechanical stirring machines consist of an electric motor clamped above the reaction vessel that rotates a vertical rod (usually glass, although it can be steel or PTFE). A vane or paddle is attached to the bottom of this rod, which is responsible for the agitation of the reaction mixture (Figure 9.32). The rod and vane are usually detachable enabling different length rods and different sized vanes to be used as appropriate. As with magnetic stirrer machines, the rate of stirring can usually be adjusted by means of a control on the motor. There are many different designs for the vane, the most common being a crescent-shaped piece of PTFE about 5 mm thick. This has a slot in it that allows easy attachment to the glass rod (Figure 9.33). In this design, the vane can be rotated about a horizontal axis and so can easily be put through the narrow neck of a round-bottomed flask and then rotated into a horizontal position ready for use.

Because the mechanical stirrer requires a physical attachment to the reaction flask, precautions have to be taken if the reaction is to be carried

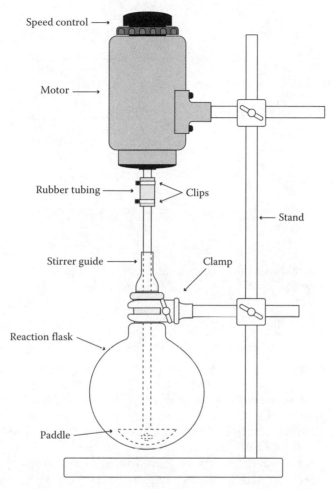

Figure 9.32 Using a mechanical stirrer.

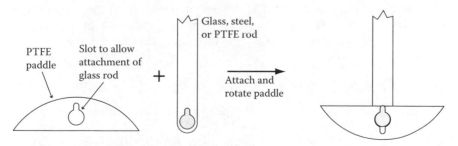

Figure 9.33 Attaching a PTFE paddle.

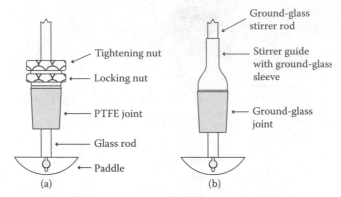

Figure 9.34 Apparatus for attaching a stirrer rod to a reaction flask: (a) PTFE joint and (b) ground-glass joint.

out under anhydrous conditions or under an inert gas atmosphere. The usual way of doing this is to use a stirrer guide that allows the rod to enter the reaction flask, but prevents atmospheric gases from doing so. These are typically constructed from PTFE (Figure 9.34a) and provide a tight fit between a ground glass joint and the glass rod. The tight seal around the rod is achieved by means of an O-ring seal inside the stirrer guide. A setup of this type should be good enough to withstand a vacuum of 0.5–0.1 mmHg inside the reaction flask without leaking. Because the guide is constructed entirely from PTFE, it does not require lubrication.

Such guides are necessarily expensive, and for many uses that do not require rigorously controlled reaction conditions, a glass guide can be employed (Figure 9.34b). This consists of a precision ground glass tube attached to a normal ground glass joint. In this case, it is essential to use a ground glass rod that forms a good fit with the tube. Oil lubrication is required to allow the rod to rotate, and so, this setup is not suitable for reactions above room temperature that involve volatile solvents that can leach the lubricant into the reaction mixture.

Because mechanical stirrers use an electric motor mounted above the reaction flask, there is a potential danger of sparks from the motor igniting volatile flammable solvents. In such instances, the use of mechanical stirrers powered by compressed air is recommended.

9.6.3 Mechanical shakers and vortexers

Mechanical shakers come in many designs and are simply motors that will shake an attached reaction flask. The flask is usually clamped to the shaker (Figure 9.35), often with a counterweight to balance the machine.

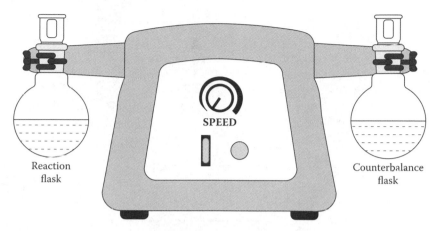

Figure 9.35 Mechanical shaker.

This is a useful device for reactions that involve prolonged vigorous mixing of two immiscible liquids and can also be employed when efficient mixing of a gas and a liquid or solid is required (e.g., hydrogenation). For most applications, other methods of agitation are preferable, but shaking can be useful as it typically generates a lower level of mechanical impact than magnetic stirring. Vortexing is a variation on shaking that is now becoming commonplace. It is often employed for biocatalytic transformations and processes involving solid-supported reagents, where mechanical impact associated with stirring methods can be problematic. Vortexers are also commonly employed in multi-parallel reactors, where it is often not possible to have individual magnetic or mechanical stirrers.

9.6.4 Sonication

Ultrasonic waves can also be employed as a means of agitation. The most common arrangement is to use a simple ultrasonic cleaning bath, in which the reaction vessel is placed (Figure 9.36), although ultrasonic probes can also be used and can be placed inside a bath or the reaction vessel (Figure 9.37).

The latter arrangement is particularly desirable if precise control of the ultrasound frequency is required or if external control of the reaction temperature is necessary. In both cases, ultrasonic waves are generated inside the reaction vessel, causing agitation of its contents. This technique is particularly useful for reactions involving insoluble solids. The ultrasonic waves break up the solids into very small particles facilitating solvolysis and reaction.

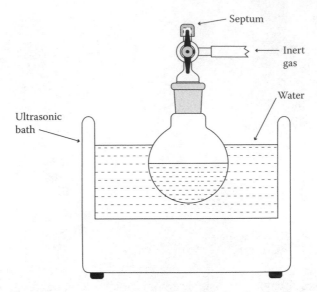

Figure 9.36 Performing a reaction in an ultrasonic cleaning bath.

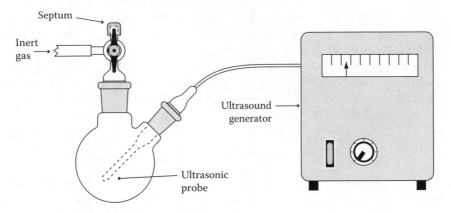

Figure 9.37 Using an ultrasonic probe.

9.7   Use of controlled reactor systems

A key skill for a chemist is to be confident about the exact conditions under which an experiment is carried out. Sections 9.1, 9.2, 9.4, 9.5, and 9.6 of this chapter provide advice on how to carry out reactions under controlled, reproducible conditions using standard laboratory equipment. There are also a range of automated integrated reactor devices available that enable chemists to get precise control over reaction conditions more easily. When using these devices, the same basic principles apply for setting up and carrying out reactions. In Section 9.7.1, some advice will be given on use of equipment for carrying out single reactions, and in Section 9.7.2, some guidance is provided for working with parallel reactor systems.

Precise control of conditions is particularly important for large-scale reactions and this is discussed further in Chapter 13. Careful monitoring of the reaction parameters on a small scale will usually provide a good understanding of the reaction characteristics allowing subsequent optimization and/or scale-up. A number of manufacturers now offer quite sophisticated reaction vessels with integrated temperature control and controlled addition of reagents. The temperature of these reactors is normally controlled either by a fluid circulator or by Peltier devices. Some of the integrated systems are designed to enable precise calorimetric measurements providing detailed information on the thermal characteristics of a reaction.

9.7.1   Jacketed vessels

Some systems employ a simple jacketed vessel (Figure 4.15) connected to a heater–chiller circulator and allow you to build a range of reaction setups for carrying out controlled reactions at subambient, ambient, or elevated temperatures. The main difference between using a jacketed vessel and using a round-bottom flask is that the jacketed vessel is normally clamped in a rigid configuration. This means that all reagents and solvents need to be measured/weighed remotely and then added to the vessel. Most jacketed vessels have a bottom runoff valve, and this conveniently allows them to be used as a separating funnel during reaction work-up. Mechanical stirring tends to be used for agitation as it is generally more controllable and reliable than magnetic stirring. Jacketed vessels can be dried by evacuating and then heating using the heater–chiller circulator. They can then be placed under inert atmosphere by flushing with nitrogen or argon. Pre-weighed solid reactants can be added using a solid addition funnel. Measured quantities of solvent can be added to the vessel in the same way as for any other reaction flask, and liquid reagents can be added in a controlled manner via an addition funnel or syringe pump. Typical setups for using jacketed vessels at both elevated and low temperatures are shown in Chapter 13.

The peripheral electronic devices used with a jacketed vessel (e.g., internal digital thermometer, heater–chiller circulator, and syringe pump) can usually be programmed independently, but often they can also be integrated so as to allow additional levels of control. For example, by linking the output from a digital thermometer to a syringe pump, the rate of addition of a reagent to an exothermic reaction can be adjusted so as to maintain a particular reaction temperature. Similarly, a pH probe can be used to control the rate of addition so as to maintain a specific pH in the reaction mixture.

There are a number of different ways the reaction can be worked up in a jacketed vessel. Adding an aqueous work-up mixture to the vessel followed by phase separation is very easy because the vessel design allows the lower phase to be drained off. Alternatively, the entire mixture can simply be drained off and worked up outside the vessel. One advantage of carrying out aqueous separations within the vessel is that they can be done at a controlled temperature. For example, a low-temperature extraction might avoid product hydrolysis/decomposition, or an elevated-temperature extraction may be useful for a product that is not very soluble. If it is necessary to quench a reaction mixture *into* another liquid (usually aqueous) at a controlled temperature, this can be done simply by transferring the reaction mixture to a second jacketed vessel containing the work-up mixture.

Jacketed vessels can also be useful for carrying out controlled crystallization, with or without an antisolvent. In a typical procedure, the compound that is to be crystallized is dissolved in the minimum amount of solvent at elevated temperature in the jacketed vessel. Supersaturation is then brought about by controlled (slow) cooling of the mixture, using the heater–chiller circulator and/or by controlled addition of an antisolvent via a syringe pump. Further control over crystal growth can be achieved by programmed temperature ramping following nucleation of the crystallization.

9.7.2 Parallel reactors

When considering a new reaction, it is tempting to rush into the laboratory and try to perform the reaction as soon as possible, but it is worth spending some time considering and planning the appropriate experiment(s) to do before you start. Instead of performing a single experiment, it is often better to set up a series of parallel experiments to identify the optimal reaction conditions reagents. Traditionally, this is done by varying individual reaction parameters (e.g., reagent, solvent, concentration, and temperature) one at a time. However, this can be done in a more systematic way using "DoE." DoE is a statistical approach to experimental design that enables you to identify a set of experiments that will provide better information than you would be able to accumulate by carrying out one experiment at time. Changing the experimental parameters simultaneously in a planned statistical fashion can reveal interactions between parameters that will not be apparent if you optimize a reaction using traditional approaches. It can also be more efficient, often allowing reaction optimization using as few as a quarter of the number of reactions required for the traditional approach. A detailed explanation of the DoE approach is beyond the scope of this book, but a range of useful references is provided at the end of this chapter.[5-8]

Once you have designed a series of parallel experiments, it is very important to plan carefully how you intend to carry them out. A lot of time and effort can be wasted if the experiments are started without thinking through the procedures and how you are going to analyze the outcome to get the information you want. Be particularly careful about quantitative analysis. For example, if you are using HPLC, the response factors for different molecules will be different. This may not matter if you are simply looking for, say, the disappearance of starting material, but if you require accurate quantification, you will need to take the response factors into account or use an internal standard (Section 9.3.2). The scale that you work on is also an important factor. It is generally easier to get reproducible results from larger-scale reactions, but when you have a limited amount of material available or need to try a large number of reagents and/or reaction conditions, small-scale reactions (<1 cm^3) may be necessary. For reactions that require dry and/or inert conditions, think carefully about how this can be done (see Section 9.2). Remember that on a small scale, even tiny amounts of water will have a relatively large impact on anhydrous reactions, so great care is needed.

A wide variety of equipment designed to facilitate parallel experimentation are now available. Some of the most commonly used items are described in Section 4.5, but the type of equipment you use may depend on what is available in your laboratory and/or your budget. When you are working on a large series of experiments, weighing and measuring reagents and reactants can be tedious, and accuracy can be challenging when working on a small scale. Modern weighing robots are very accurate and useful for weighing reagents into tubes. If a robot is not available, it is often best to make up stock solutions of reactants and reagents. This allows for easy measurement and transfer of small quantities of material using a syringe or (multi)pipette. Alternatively, automated liquid handling equipment can be used to dose each reaction with a required amount according to a programmed regime.

If you need to perform a series of reactions at different temperatures, it is best to use a multi-parallel reactor that incorporates a heater–chiller unit. The most sophisticated of these allow each reaction to be performed at a different temperature and provide calorimetric data on each reaction. More basic devices only allow the reactions to be performed at a single temperature.

Agitation in parallel reactor tubes can be problematic and is a common cause of unexpected results. It is important to select the size and type of stirrer carefully, so as to maintain efficient mixing throughout the reaction. Particular care should be taken when using long thin reaction tubes as it is tempting to fill them to a height where stirring becomes ineffective.

Parallel reactions are commonly used to collect and compare reaction profile data. In many instances, this requires each reaction to be sampled at regular intervals. This can be a tedious process unless the sampling is automated. Fortunately, there is a range of simple automated liquid sampling devices available for tube reactors. Typically, these can be programmed to take samples from each reaction at set intervals and carry out simple reaction quench or work-up if required. Usually, they will also transfer a portion of the sample to in-line analytical instrumentation (e.g., HPLC and GC) to complete the analysis. This type of equipment allows you to set up a series of parallel reactions and to leave them running, knowing that a set of samples will be collected and analyzed. As some of the sample is retained, this can also be reanalyzed later using any method of choice. Of course, samples can also be taken manually if autosampling equipment is not available. Rapid HPLC/UPLC or LC-MS systems are available for analyzing reactions as they progress, but it is often convenient to put all the samples into an LC or LC-MS autosampler once the reaction set is complete. TLC does not provide the same level of quantification as HPLC/UPLC, but it can still be an effective way to follow parallel reactions if more sophisticated analytical instrumentation is not available. If TLC is used, it is preferable to run multiple samples on the same plate, similar to the way in which flash chromatography is monitored (Figure 11.20), as this makes it much easier to spot differences between reactions. An exciting recent innovation is the development of simple MS sampling ionization techniques that allow mass spectra to be obtained directly from a spot taken from a reaction mixture. The data obtained can be a simple spread of ions, ion current for a single ion, or high-resolution ion measurements, depending on the type of mass spectrometer used. Miniaturized mass spectrometers are also available, allowing the instrument to be brought to the experiment.

An alternative to sampling the reactions is to use direct analysis. In its simplest form, this can be measurement of physical parameters such as heat output, pH change, turbidity, or gas evolution, all of which can be used in conjunction with the more sophisticated multi-parallel reactors. Some reactors can also be used with spectroscopic monitoring using, for example, IR, UV, or Raman, but for these techniques, standards need to be prepared and analyzed to be able to interpret the data generated.

From the preceding discussion, it should be clear that carrying out multi-parallel experiments can enable you to obtain a detailed understanding of a reaction quickly and efficiently. The apparatus used can range from a series of vials in an aluminum block to very sophisticated controlled reactors, depending on what data you require and what equipment you have in your laboratory. In all cases, the most important thing is to plan the execution and analysis of the experiments carefully, so that you collect the data that you need with an appropriate level of accuracy.

References

1. A.J. Gordon and R.A. Ford, *The Chemist's Companion*, J. Wiley and Sons, New York, 1972.
2. A.M. Phillips and D.N. Hume, *J. Chem. Ed.*, 1968, **54**, 664.
3. R.E. Rondeau, *J. Chem. Eng. Data*, 1966, **11**, 124.
4. N.S. Isaacs and A.V. George, *Chem. Brit.*, 1987, **23**, 47.
5. R. Carlson and J.E. Carlson, *Design and Optimization in Organic Synthesis*, 2nd ed., Elsevier, Amsterdam, 2005.
6. M.J. Anderson and P.J. Whitcomb, *DOE Simplified: Practical Tools for Effective Experimentation*, 2nd ed., Productivity Press, New York, 2007.
7. M.J. Anderson and P.J. Whitcomb, *RSM Simplified: Optimizing Processes Using Response Surface Methods for Design of Experiments*, Productivity Press, New York, 2004.
8. L. Ericksson, E. Johansson, N. Kettaneh-Wold, C. Wikstrom, and S. Wold, *Design of Experiments, Principles and Applications*, 3rd ed., Umetrics Academy, Umeå, Sweden, 2008.

chapter ten

Working up the reaction

10.1 Introduction

It is important to give some thought to the work-up of the reaction before you attempt it. There are several aspects that need to be considered, the most important of which are as follows: First of all, make sure that the reaction has actually finished (by careful analysis using your chosen monitoring system). When using TLC analysis, it is sometimes difficult to judge by spotting the reaction mixture directly on to the TLC plate. This is because other components in the reaction mixture can sometimes obscure the spots of interest. High-boiling solvents such as DMF or pyridine can also obscure TLC spots of interest. If you suspect this is a problem, it is often possible to get a more accurate TLC analysis by withdrawing a small aliquot of reaction mixture by syringe and performing a "mini work-up" in a small vial. To do this, add your aliquot to a vial containing 0.5 cm^3 each of diethyl ether and an appropriate aqueous solution. Shake the vial vigorously, then TLC the diethyl ether layer. This technique can also be used to screen alternative work-up conditions, for example, adding to water or an aqueous base rather than aqueous acid, or using other organic solvents in place of diethyl ether.

Having satisfied yourself that the reaction has run to completion or that it is time to end the experiment, the appropriate "quench" is added to the reaction mixture. Choice of this reagent can be crucial and can greatly affect the isolated yield of desired product. It is also vital to use a quench reagent or procedure that is safe. If the product is expected to be reasonably stable, which usually is the case, then the choice of procedure for quenching the reaction is determined by the reagent(s) used in the reaction. Clearly, we cannot cover all possibilities in this chapter, but general procedures that should cover most of the situations that are likely to be encountered are provided in Section 10.2. The classification is made on the basis of the nature of the reaction mixture that is to be worked up.

10.2 Quenching the reaction

If the reaction has been carried out under an inert atmosphere, then it is advisable to add the quench before exposing the reaction mixture to the air. It is best added as you would add a reagent (typically dropwise by syringe). If the reaction was run at low temperature, then add the quench at this

temperature and allow to warm to room temperature slowly, before opening to the air and proceeding with the isolation of the product. This should
ensure that the outcome of the reaction is a true reflection of what occurred
at low temperature. Do not forget that most aqueous solutions will freeze
below −10°C and therefore will not necessarily quench a reaction that is
substantially cooler than this. If you need to be sure that a highly basic
reagent has been quenched at low temperature, it is often more effective
to do this with an equivalent of methanol or acetic acid rather than water.

If the reaction was run at elevated temperature, allow it to cool to room
temperature before adding the quench (still under the inert atmosphere,
if used). If an exotherm is possible, ensure that the reaction is cooled in a
cooling bath (e.g., ice/water) and that the quench is added very carefully
(if the scale of the reaction is such that an internal thermometer can be
used, then do so and keep a close eye on the internal temperature while
quenching the reaction).

These general comments apply to the following specific types of reactions. Obviously, if a known, safe, and fully described literature procedure is being used, then follow the work-up and isolation process exactly.
Modify such a procedure only if you encounter problems.

10.2.1 Strongly basic nonaqueous reactions

This is a commonly encountered situation in organic synthesis; typical
examples would include processes using strong bases (e.g., n-BuLi,
(i-Pr)$_2$NLi) or reactive organometallics (e.g., MeLi, Grignard reagents, and
cuprates). The most commonly used quench for this type of reaction is to
cautiously add an excess of aqueous ammonium chloride. This will protonate any anion present and quench the excess reagent. The quench can
be added to reactions at low temperature, but if you are concerned about
the aqueous solution freezing out, then use an equivalent acetic acid or
methanol as a prequench before adding the ammonium chloride solution.

10.2.2 Near neutral nonaqueous reactions

This situation covers most reactions that do not involve a strong acid or
a strong base. It would normally include reactions that are catalyzed by
weak acids (e.g., acetal formation) or weak bases (e.g., Michael addition).
Clearly, a great many reactions fall into this category and in general, the
quench can be aqueous ammonium chloride for mildly basic reactions
and aqueous sodium hydrogen carbonate for mildly acidic reactions.

If a fairly reactive reagent has been used, then add the quench carefully with cooling. Sometimes, it may also be beneficial to stir the quench
for a while to ensure destruction of excess reagents. For example, if an
acid anhydride has been used to prepare an ester, it is often useful to add

a relatively small volume of aqueous sodium hydrogen carbonate and stir for an hour or so to destroy any remaining anhydride. This will form the sodium carboxylate that will then remain in the aqueous phase during the subsequent extraction.

As always, it is important to know whether your product is likely to be sensitive to (or destroyed by) the quench. For example, any reaction that has been driven to completion by removal of water needs some thought. A common example of this is an acid-catalyzed enamine formation. The product enamine is likely to be quite sensitive to aqueous acids, so quench with aqueous sodium hydrogen carbonate and make sure that the product is in contact with the moisture only for a short period.

For products that are particularly sensitive to acid or base, an aqueous buffer of the appropriate pH can be used. If an amine (e.g., triethylamine or pyridine) has been used in the reaction, it is usually removed by washing with dilute acid. If the product is acid sensitive, then washing with aqueous copper sulfate will usually remove the amine.

10.2.3   Strongly acidic nonaqueous reactions

This type of reaction will typically involve the use of a strong Lewis acid such as $TiCl_4$ or $BF_3.OEt_2$. It is important to be aware that the addition of water to these reagents will be exothermic and is likely to liberate a strong protic acid. If the product is likely to be unaffected by the acid liberated, then water can be used as a quench. However, it is often the case that the product will be unstable toward strong protic acid or at least contain functionality that will be affected by the acid. In this case, aqueous sodium bicarbonate or aqueous sodium carbonate can be used. If a nonaqueous quench is desired, then a solution of gaseous ammonia in the reaction solvent can be used. Quench these reactions carefully, cooling. Be aware that some metal salts, Al(III) especially, can give rise to emulsions or gelatinous precipitates, and be prepared to deal with the problems that this might cause.

10.2.4   Nonaqueous reactions involving Al(III) reagents

Aluminum (III)-based reagents are commonly used in organic synthesis. In particular, many reactive hydride reducing agents are based on Al(III), such as $LiAlH_4$ and $(i\text{-}Bu)_2AlH$. With these reagents, a simple aqueous acid work-up can often lead to emulsions or gel formation, making it difficult to extract the product. If this occurs, an alternative procedure that often gives good results is to add a saturated aqueous solution of sodium sulfate dropwise with stirring (and cooling!), until a very heavy (sodium alum) precipitate is formed. For best results, just add enough aqueous sodium sulfate to incorporate all of the aluminum in the alum structure. The alum formed has the formula $NaAl(SO_4)_2.12H_2O$, so the exact quantity of aqueous sodium sulfate required can be calculated. The organic supernatant can then be decanted and the alum residue is extracted a few times with organic solvent to ensure recovery of your entire product. Do not dispose of the solid until you are certain that you have obtained a good recovery of material; it might be necessary to extract further to obtain the entire product.

An alternative work-up procedure specifically for LiAlH$_4$ is as follows. The stirred reaction mixture is first diluted with diethyl ether, then the following are added dropwise (*caution!*), in sequence: (1) water (1 cm^3 per gram of LiAlH$_4$ used), (2) aqueous sodium hydroxide (15% solution and 1 cm^3 per gram of LiAlH$_4$ used), and (3) water (3 cm^3 per gram of LiAlH$_4$ used). This will often produce a granular precipitate, which is easy to filter and wash.

Even with these precautions, you may end up with significant amounts of your product sequestered in the aluminum residues. This problem is often encountered because hydride reductions are commonly used to generate very polar organic products (e.g., amino alcohols and polyhydroxylated compounds). In these situations, it is often possible to extract more products by using a Soxhlet extractor. This comprises a Soxhlet apparatus (Figure 10.1a) and a filter thimble (Figure 10.1b). This apparatus allows continuous extraction of a solid with hot solvent, and it can be very effective for isolating polar reaction products from inorganic solids.

To perform a Soxhlet extraction, you first place the solid in the filter thimble. It is important to use an appropriately sized thimble and not to fill the thimble more than half full. The thimble is then placed inside the Soxhlet apparatus. Make sure that the sides of the thimble are higher than the top of the siphon, otherwise the solid will be washed out of the thimble during the extraction process.

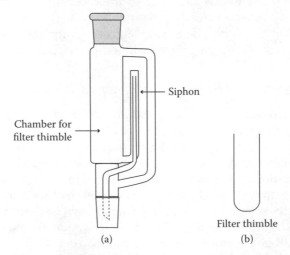

Siphon

Chamber for
filter thimble

Filter thimble

(a) (b)

Figure 10.1 Soxhlet apparatus.

The Soxhlet apparatus is then attached to a round-bottom flask containing the extraction solvent and fitted with a condenser (Figure 10.2). When the solvent is heated at reflux, it condenses into the chamber containing the filter thimble, filling the chamber with hot solvent. As soon as the solvent level reaches the top of the siphon, it pours back into the flask, emptying the chamber. This process will be repeated as long as you leave the apparatus running and so is equivalent to extracting the solid with hot solvent many times. Once the extraction is complete, you simply allow the apparatus to cool to room temperature, dismantle, and evaporate the solvent in the round-bottom flask to recover your product.

10.2.5   Reactions involving oxidizing mixtures that may contain peroxide residues

Great care should be taken when working up oxidation reactions where peroxide residues might be present at the end of the reaction. Common examples are reactions using hydrogen peroxide, a hydroperoxide (e.g., tert-butyl hydroperoxide), or a peracid. In such instances, the work-up should always include treatment with an appropriate reducing agent such as sodium metabisulfite or sodium thiosulfate. A simple aqueous wash with a solution of these reagents will often suffice, but in some instances, it may be necessary to stir the reaction mixture with the reducing agent for a while. If the product of the reaction does not tolerate an aqueous work-up, the reducing agent can be used as a part of a solid–liquid work-up, as described in Section 10.3.5. After carrying out the reductive work-up, it is essential to test the mixture to ensure there is no residual peroxide before concentrating or heating the crude product mixture. Peroxide test strips are a convenient way to test for the presence of peroxides.

10.2.6   Acidic or basic aqueous reactions

To quench these reactions, it is usually enough to neutralize with dilute aqueous acids or bases. You can then extract the product into an organic solvent or collect by filtration if it precipitates from the solution.

10.2.7   Liquid ammonia reactions

A number of synthetically useful reactions use liquid ammonia as a solvent. Most involve a generation of strongly basic species. The usual quench for this type of reaction is to carefully add the excess solid ammonium chloride and then allow the ammonia (boiling point [b.p.]: −33°C) to evaporate (see also Section 14.6).

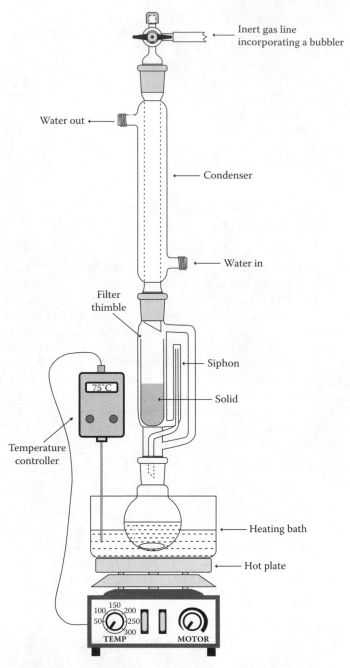

Figure 10.2 Soxhlet extraction.

10.2.8   Reactions involving homogeneous transition metal catalysts

Homogeneous transition metal catalysts are widely used in organic synthesis. The most prominent are Pd-catalyzed processes, such as the Suzuki, Buchwald, and Heck couplings. These processes typically employ a metal catalyst that is rendered soluble in organic solvents via complexation with organic ligands. Although these catalysts are used in substoichiometric quantities (typically 1–5 mol%), the metal + ligand(s) often have a high molecular mass, so the amount of material by weight can be quite high. Insoluble metal residues (e.g., Pd-black) formed during the reaction can usually be filtered off using a filter aid such as Celite. However, removal of the remaining soluble catalyst can be quite challenging. Along with the catalyst, there may also be stoichiometric reagents such as a base, and these may be converted into a different substance during the reaction. Consequently, understanding the fate of all the components that went into the reaction is crucial when designing the work-up procedure.

An example of a palladium catalyzed amine coupling reaction is shown in Figure 10.3. This illustrates the range of substances that may be present during the work-up of a reaction involving a homogeneous transition metal catalyst. In this process, a range of ligated Pd species remain in solution at the end of the reaction and are not easily separated from the reaction product.

The work-up procedure that was devised for this reaction illustrates how several techniques can be combined to good effect. First, the reaction mixture was filtered through Celite to remove traces of palladium metal and the insoluble inorganic salts (mainly CsI and Cs_2CO_3). All other components of the mixture remained in the solution and proved very difficult to separate using standard techniques. In this case, the solution was to treat the mixture with methanolic sodium hydroxide. This converted the ester group in the product to the corresponding carboxylic acid sodium salt. This was then far less soluble than the remaining catalyst and by-products and so readily crystallized from the solution. In this case, the

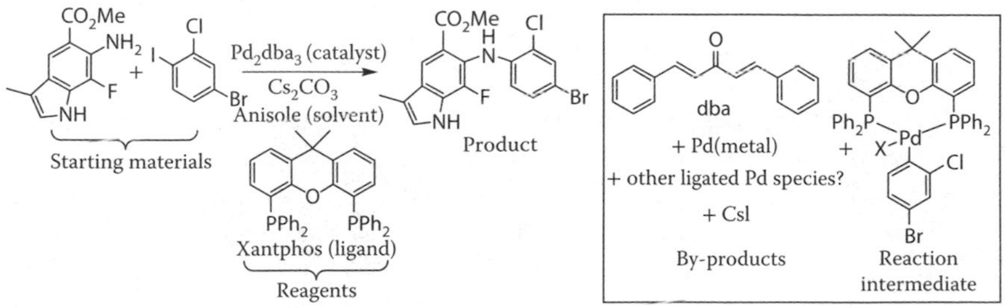

Figure 10.3 Example of a Pd-mediated coupling.

Pd content in the isolated sodium salt was low, but persistent heavy metal residues can sometimes be a problem, which may require removal using a scavenger (see Section 10.3.7).

This example illustrates how a simple transformation can be used to dramatically change the physical properties of one component, facilitating its isolation/removal from a complex reaction mixture. In this case, it was the product that was altered, but in other processes, it may be more appropriate to chemically alter the by-product(s), such that they can be more readily removed from a reaction mixture.

10.3   Isolation of the crude product

Once the quench is complete, the next stage is to isolate your crude product from the mixture. If chosen correctly, the isolation procedure can also result in partial purification of the desired product and hence simplify any subsequent purification process.

For solid products, direct crystallization from the reaction mixture should be considered as a potential work-up and/or purification technique. It is often possible to devise reaction conditions where the product is less soluble than the starting material(s) and reagents. In such cases, careful choice of solvent, concentration, and temperature can often lead to direct crystallization of the product from the reaction mixture. It can then be collected by filtration. Alternatively, it may be possible to add an antisolvent at the end of the reaction to selectively precipitate the product.

If the reaction product is either acidic or basic, it may be possible to crystallize it from an aqueous reaction mixture by neutralization. Basic solids will precipitate from aqueous acids on neutralization and acidic solids will precipitate from aqueous bases on neutralization.

Obtaining good quality crystals is often the key to avoid impurities in the crystalline solid. Careful choice of the crystallization solvent(s), temperature, and concentration is the key to this.

If direct crystallization from a reaction mixture is not possible, you may be able to crystallize the product from the organic extract after performing an aqueous work-up. Again the key to a successful crystallization that excludes impurities is good choice of solvent(s), concentration, and temperature. An antisolvent will be needed here if the product is too soluble in the primary extraction solvent.

If the product still contains impurities after precipitation, it can often be improved by washing during the filtration process. Typically, this is done by slurrying the product filter cake with a small amount of solvent that dissolves the impurities more readily than the product. The solvent is then sucked off and the process is repeated as necessary. Care should be taken when using this technique, because it is easy to lose significant amounts of product if it is partially soluble in the solvent that is being used.

10.3.1 Typical isolation from an aqueous work-up

The most common type of work-up employed in organic chemistry involves partitioning the reaction mixture between an immiscible organic solvent and an aqueous solution. Before carrying this out, it is sometimes necessary to remove any insoluble solids that are present in the reaction mixture. This is best achieved by first diluting the (quenched) reaction mixture with organic solvent and then vacuum filtering through a Celite pad in a sintered glass funnel (Figure 10.4). Rinse through two or three times with more organic solvent to ensure complete transfer of your product.

If your organic reaction solvent is miscible with water (e.g., methanol), then it is advisable to remove as much of it as possible using a rotary evaporator, prior to extraction with a water-immiscible organic solvent. This is not always essential, but it will usually make the aqueous extraction more straightforward. This will also work for solvents that are partially miscible with water (e.g., tetrahydrofuran [THF]), although here it may also be possible to separate the solvent completely, simply by saturating the aqueous phase with NaCl.

The reaction mixture can then be partitioned between an aqueous solution and an immiscible organic solvent. This is most commonly done using a separating funnel (Figure 10.5). Because the two solutions are immiscible, they form two separate layers. The bottom layer can then easily be separated by draining through the two-way tap.

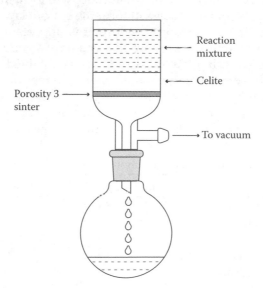

Figure 10.4 Filtration through Celite® to remove insoluble solids.

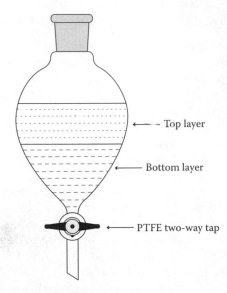

Figure 10.5 Using a separating funnel.

In general, nonpolar organic molecules will be retained in the organic phase, and very polar organic molecules and inorganic salts will be transferred to the aqueous phase. However, the degree to which this occurs is dependent on the exact nature of the organic and aqueous phases. The most common organic solvents used in aqueous extractions are listed in Table 10.1. Those that are less dense than water will form the top layer and those that are denser than water will form the bottom layer.

Throughout the extraction procedure, it is important not to discard any of the extracts until you are sure that you have a good mass recovery of the crude product. If a low mass recovery is obtained, the rest of the material must be somewhere and could well be in the aqueous phase. For most reactions, ethyl acetate will serve as the best extraction solvent. It is less dense than water and so will normally be the top layer in a separating funnel. However, if this results in a low recovery of product, try saturating

Table 10.1 Properties of Commonly Used Extraction Solvents

Solvent	Boiling point (°C)	Density (g/cm³)
Diethyl ether	35	0.71
Ethyl acetate	77	0.90
Water	100	1.00
Saturated aqueous NaCl	109	1.20
Dichloromethane	40	1.32
Chloroform	61	1.49

the aqueous layer with sodium chloride (to increase the ionic strength), and/or extracting with chloroform. Whenever you extract with an organic solvent, always use several small portions rather than one large one as this is much more efficient.

Occasionally, you will encounter the problem of emulsions during extraction. There is no universal solution to this, but a common cause is the presence of very fine particles, which can be removed by filtration through Celite as described above. If this fails, try adding sodium chloride to increase the density of the aqueous phase or diethyl ether to decrease the density of the organic phase.

If simple extraction with a single organic solvent fails, there are a number of options available. You could concentrate the aqueous layer under reduced pressure and then extract the residue with organic solvent. In extreme cases, this residue could be extracted using a Soxhlet apparatus (Figure 10.2). If your product is unstable to the concentration in aqueous solutions, then extraction of the aqueous (after saturation with sodium chloride) with a mixture of chloroform and ethanol (2:1) is sometimes successful. If this fails, try continuous extraction of the aqueous using a Hershberg–Wolfe apparatus (Figure 10.6). This apparatus works best if an organic solvent with a higher density than water (e.g., chloroform) is used. The solvent is placed in a round-bottom flask and heated to reflux. The hot solvent then distils into the Hershberg–Wolfe apparatus. If the solvent has a higher density than water, it will fall through the aqueous phase, extracting the product as it falls, and form a lower layer. This is then pushed back into the distillation pot by the weight of the aqueous phase. The two-way tap shown in Figure 10.6 is not essential when the Hershberg–Wolfe apparatus is used in this way. However, it is useful as it allows the apparatus to be modified for use with solvents that have a lower density than water.

If all attempts at extracting the product from an aqueous solution fail, it may be necessary to use a nonaqueous work-up (see Section 10.3.5).

Once you have combined the organic extracts containing your product, they should be dried by swirling with an insoluble inorganic drying agent. Anhydrous magnesium sulfate is most commonly used for this purpose because it rapidly takes up water and also visibly changes from a fine powder to a sticky solid when this occurs. However, anhydrous magnesium sulfate is Lewis acidic, so for acid sensitive compounds anhydrous sodium sulfate or anhydrous potassium carbonate are usually better. These tend to act slowly, so it is best to stand the organic solution over these reagents for 10 minutes to ensure optimal removal of residual water.

Finally, filtration of the drying agent, followed by evaporation of the organic solvent using a rotary evaporator will furnish the crude reaction product. Any remaining traces of solvent can be removed by placing the sample under high vacuum, but be careful, if your product is volatile (BP < 150°C), it may disappear along with the solvent.

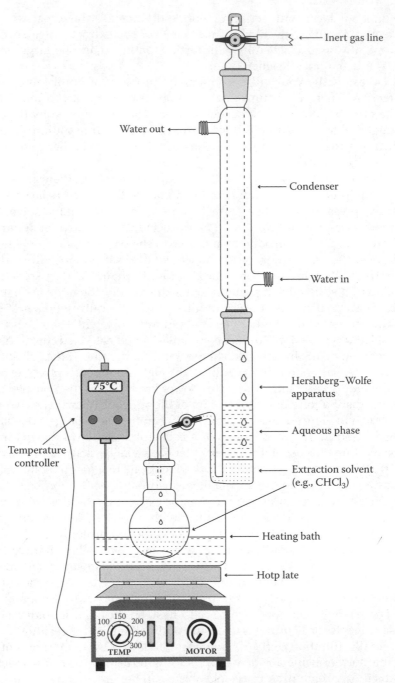

Inert gas line

Water out

Condenser

Water in

75°C

Hershberg–Wolfe
apparatus

Aqueous phase

Extraction solvent
(e.g., CHCl₃)

Temperature
controller

Heating bath

Hotp late

150
100   200
50   250
       300
TEMP   MOTOR

Figure 10.6 Continuous liquid–liquid extraction using a Hershberg–Wolfe apparatus.

10.3.2 Isolation from a reaction involving nonvolatile polar aprotic solvents

This situation is commonly encountered in organic chemistry. The most common solvents involved are DMSO and *N,N*-DMF. With these solvents, it is often possible to remove most, if not all, by adding the quenched reaction mixture to a relatively large volume of water and extracting several times with diethyl ether. The combined organics then need to be washed with water (typically five more times) to remove all the polar aprotic solvent. Any remaining traces of DMF or DMSO can then be removed during purification of the crude product.

Occasionally, HMPA may have been used as an additive. This can be extracted from an organic solution by washing with aqueous lithium chloride. HMPA is carcinogenic, so appropriate precautions should be taken when handling and disposing of aqueous waste containing this compound.

10.3.3 Using an acid/base aqueous work-up to separate neutral organics from amines

It is possible to modify a simple aqueous work-up to allow easy separation of neutral and basic organic molecules. This technique is most commonly used for the purification of low molecular weight amine products, and it can be illustrated by the example shown in Figure 10.7.

This reaction involves the hydrolysis of an imine **A** to give the desired product amine **B**, and a by-product, benzophenone **C**. The reaction is performed using an aqueous acid with THF as a co-solvent. At the end of the reaction, the amine needs to be separated from the mixture. If a mild base work-up is used, the organic extracts will contain the desired amine **B**, THF, and benzophenone **C**. These would then need to be separated to obtain the pure amine. THF is easily removed by evaporation (BP: 66°C) but benzophenone **C** (BP: 306°C) is not, so it would need to be removed by chromatography or vacuum distillation. However, because amine **A** is basic, it will protonate in the aqueous acid and hence become highly water soluble. Benzophenone **C** and THF are neutral organic molecules; consequently, they are not readily protonated with the aqueous acid. These properties can be used to separate amine **B** from the other reaction components as outlined in Figure 10.8.

$$Ph_2C=N \underset{Ph}{\overset{CO_2t\text{-}Bu}{\bigvee}} \quad \xrightarrow{THF,\ H_3O^+} \quad H_2N \underset{Ph}{\overset{CO_2t\text{-}Bu}{\bigvee}} \quad + \quad Ph_2C=O$$

A   B   C

Figure 10.7 Amine formation.

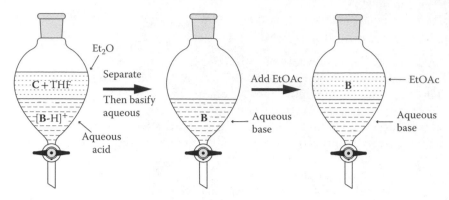

Figure 10.8 Using an acid/base work-up to purify an amine.

The reaction mixture is first partitioned between the aqueous acid and diethyl ether (Et$_2$O). At this point, the neutral organic molecules will extract into the Et$_2$O phase and the protonated amine (**B-H$^+$**) will remain in the aqueous phase. Once the Et$_2$O layer has been separated, the aqueous phase can be basified to generate the free amine **B**. This will now readily extract into an organic solvent such as ethyl acetate (EtOAc). Finally, separation of the ethyl acetate phase, followed by drying and evaporation, will give amine **B**. In many cases, this simple procedure will provide amine products of sufficient quality that further purification is not necessary.

10.3.4 Using an acid/base aqueous work-up to separate neutral organics from carboxylic acids

The procedure outlined in Section 10.3.3 can be adapted to allow separation of acidic low molecular weight organic compounds (e.g., carboxylic acids) from neutral organic by-products. A typical example of where this would be used would be in the hydrolysis of a carboxylic acid ester, as shown in Figure 10.9.

In this case, the mixture is first partitioned between an aqueous base and an organic solvent (Figure 10.10). The acidic compound will be deprotonated in an aqueous base giving a highly water soluble salt (**B$^-$Na$^+$**). The neutral organic species (THF and benzyl alcohol **C**) can therefore be separated by extraction into the organic phase. Acidification of the aqueous phase will then generate the carboxylic acid (**B-H**). Finally, extraction into an organic solvent, drying, and evaporation will give the carboxylic acid.

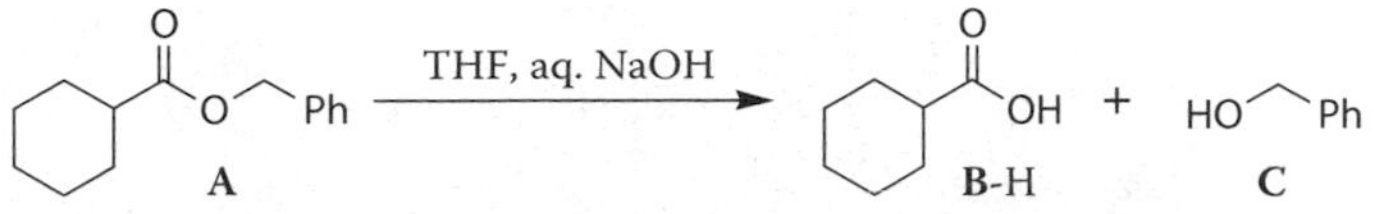

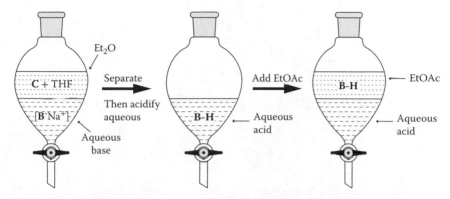

Figure 10.9 Carboxylic acid formation.

Figure 10.10 Using a base/acid work-up to purify a carboxylic acid.

10.3.5   Nonaqueous work-ups

Aqueous work-ups are a convenient way of separating organic and inorganic materials, especially when working on larger scales. However, in many cases, it is possible to isolate an organic compound from a reaction mixture without using an aqueous work-up. Indeed, on a small scale (<50 mg), it can often be more convenient not to do an aqueous work-up. Also on larger scales, it may be necessary to avoid using water due to product instability or difficulty in extracting the product from an aqueous mixture. In some instances, it may be possible to precipitate the product from the reaction mixture simply by adding an antisolvent. The product can then be collected by filtration. It is also sometimes possible to distill the product directly from the reaction mixture, although this generally works well only with highly volatile products.

If these techniques are not appropriate, a useful alternative is simply to filter the reaction mixture through layers of inorganic reagents in a sintered glass filter funnel (Figure 10.11).

Note: It is essential that you carefully quench any pyrophoric reagents prior to removing the inert atmosphere from a reaction flask.

Silica gel (as used for flash chromatography) is most commonly used because it readily absorbs most polar by-products. Other agents that are commonly used are basic alumina and charcoal, both of which are also

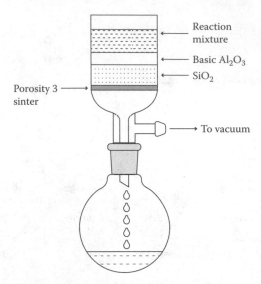

Figure 10.11 Filtration through layered reagents to remove by-products.

Table 10.2 Examples of Functionalized Silica Gel Scavengers

Functionality on silica gel	Removes
−SH	Alkyl halides, epoxides
−NHCH$_2$CH$_2$NH$_2$	H$^+$, RCOCl, RSO$_2$Cl, anhydrides
−ArSO$_2$NHNH$_2$	Aldehydes

effective for the removal of small quantities of acid from a reaction mixture. In some cases, using a combination of inorganic agents, as illustrated in Figure 10.11, may be needed to remove all the unwanted by-products. On a small scale (<50 mg), a Pasteur pipette plugged with cotton wool can be used instead of the sintered glass filter funnel (see Chapter 12).

There are also a range of functionalized silica gels (silica gel scavengers) that are now commercially available. These are silica gels that have reactive groups chemically bonded to the surface and are designed to react with specific reagents and by-products (Table 10.2). They are normally added to a reaction mixture and stirred for a period of time to allow the functionalized silica to react with any excess reagent or by-products, then removed by filtration to leave a solution of the desired reaction product.

10.3.6   Work-ups using scavenger resins

Scavenger resins can also be used as part of a nonaqueous work-up. These are most appropriate for small-scale applications and in-flow chemical

Table 10.3 Commonly Used Functionality in Scavenger Resins

Functionality	Scavenges
–SH	Alkyl halides, epoxides
–NHCH$_2$CH$_2$NH$_2$	RCOCl, RSO$_2$Cl, anhydrides, acids
–ArSO$_2$NHNH$_2$	Aldehydes
–N(CH$_2$CH$_2$OH)$_2$	Boronic acids
–N=C=O	Amines, alkoxides
–CHO	Hydrides, hydroxylamine, hydrazines
–NEt$_4$$^+HCO_3$$^-$	Acids
–ArSO$_2$Cl	Alcohols, amines

processes. Scavenger resins are insoluble polymers that are functionalized with reactive groups. The advantage of scavenger resins over functionalized silica is that a wider range of functional groups can be incorporated. The majority of these are functionalized polystyrenes and they are classified according to the reactive functionality, the degree of cross-linking of the polystyrene, and the mesh size of the resin beads. Normal polystyrene resins are cross-linked with 1%–2% divinylbenzene. These shrink on drying and only work well in solvents that swell the resin to expose the reactive functional groups buried within. Consequently, it is important to check which solvents are compatible when using these resins. Macroporous polystyrene scavenger resins are also widely available. These have a higher degree of cross-linking that makes them more rigid and hence less sensitive to shrinkage on drying. Consequently, they can be used in a wider range of solvents. They are also better for use in tightly packed columns where volume changes due to shrinkage or swelling would be a problem.

Scavenger resins can be added to a reaction mixture or used in a filtration similar to that shown in Figure 10.11. They can also be packed into columns. The reaction mixture is then worked up simply by passing it through the column. This offers the opportunity for automated work-up and purification using flow procedures.

Some examples of functionalized resins and their uses are given in Table 10.3.

10.3.7 Use of scavengers to remove heavy metal residues

Heavy metals such as palladium, platinum, rhodium, and so on are commonly used as catalysts in organic synthesis. If the metal is used on a solid support (e.g., palladium on carbon), it can usually be removed by filtration through Celite. However, if the metal is part of a homogeneous catalyst, it is often difficult to remove at the end of the reaction. This is especially true if the reaction product has functional groups (e.g., nitrogen and sulfur) that can coordinate to the metal. On a small scale, filtration through silica or flash chromatography may be effective in removing the residual metal, but this is not always the case, and on a larger scale, chromatographic purification may not be desirable. Isolation and purification of the product by crystallization is effective in removing metal residues that are retained in solution, but will not remove metal residues that are bound to the product. In such cases, scavenging the metal may be the only effective way to remove it. The key to success is to find a scavenger that will bind more tightly to the metal than your product. Finding a suitable scavenger is often a matter of trial and error, but there is now a considerable amount of guidance available in the literature and from scavenger manufacturers. The simplest and cheapest way to remove heavy metal residues is to wash the product with a solution of a scavenger, such as thiourea, or a thiourea derivative. If this is not effective, a solid-supported scavenger can be used and a wide selection of these are available.

10.4 Data that need to be collected on the crude product prior to purification

It is good practice to collect a set of analytical data (typically ^{1}H NMR, IR, mass spectrum, and TLC) on the crude product prior to purification. This is particularly important if the reaction has not been performed before. *At the very minimum*, you should collect enough data to determine whether the desired product has been formed. There is little point in attempting to purify a product that is not there. Ideally, you should also collect enough analytical data so as to determine the identity and amount of any by-products. This will enable an effective comparison of different reaction conditions during reaction optimization. It is also essential to weigh the crude product to record the mass recovery from the reaction. If this is low, you may need to redesign the work-up procedure.

chapter eleven

Purification

11.1 Introduction

When a crude product has been isolated from a reaction work-up, the next step is to purify it. The degree of purity required will depend to some extent on the use for which the sample is intended. A synthetic intermediate might only require rough purification, whereas a product for elemental analysis would require rigorous purification. This chapter describes the most important purification, crystallization, distillation, sublimation, and chromatography techniques. It is assumed that the reader is familiar with the basic principles of these methods, so the emphasis here is on more demanding applications such as the purification of air-sensitive materials and purifications on a microscale.

11.2 Crystallization

11.2.1 Simple crystallization

All crystallizations are based on the same principle: a solid compound is fully dissolved in a solvent (or solvents) at a particular temperature, the conditions of the solution are changed such that the solution becomes supersaturated in the compound to be crystallized, and the compound then crystallizes from the solution. Many compounds can crystallize in more than one polymorphic form. For example, snow and hail are different polymorphs of ice. The polymorph you get depends on the conditions under which the crystallization takes place. Polymorphs have different melting points and stabilities, so if you have difficulty in obtaining good crystals from one crystallization system it is worth trying other systems.

There are two distinct phases of crystallization:

1. Nucleation: This is when a solid first starts to precipitate from a solution. The formation of a crystalline solid rather than oil or wax can be highly solvent dependent, so the correct choice of solvent is crucial.
2. Crystal growth: This is the second phase of crystallization. In general, the slower the rate of crystallization the larger the size of the crystals formed.

There are several different ways to make a solution supersaturated in order to induce crystallization. The simplest way is to generate a hot saturated solution by dissolving the solid of interest in the minimum amount of a single solvent and then induce crystallization by cooling. Where this procedure works it is generally the preferred method. Compounds are usually far more soluble in hot solvents than in cold solvents. As a rough guide, the solubility of a compound will approximately double for every 20°C increase in temperature. For this reason, it is generally easier to find a single crystallization solvent for high-melting-point compounds because higher boiling solvents can be used. If crystallization from a single solvent is unsuccessful, an antisolvent (a solvent in which the compound is less soluble), with or without cooling, is often added. Evaporative crystallization in which supersaturation is induced by partial evaporation of the solvent is also a commonly used technique. In this case, the solution can be allowed to evaporate until crystallization occurs, or a portion of the solvent can be distilled off and crystallization is then induced by cooling. Evaporative crystallization can be performed using solvent mixtures as long as the solvent in which the compound is more soluble has the lower boiling point.

The method you choose for crystallization may depend on whether you want to maximize the yield from the crystallization or obtain a small amount of high-quality crystalline material, for example, to use for x-ray crystallography. High supersaturation will maximize yield, whereas a slow crystallization from a mildly supersaturated solution will generally provide crystals that are larger and of higher purity.

The basic procedure for simple crystallizations can be broken down into five steps:

1. Select a suitable solvent or solvent mixture. Find a suitable solvent by carrying out small-scale tests. Remember that "like dissolves like." The most commonly used solvents in order of increasing polarity are petroleum ether, toluene, acetone, ethyl acetate, ethanol, and water. It is preferable to use a solvent with a boiling point in excess of 60°C, but the boiling point should be at least 10°C lower than the melting point of the compound being crystallized in order to prevent the solute from "oiling out" of the solution. Alcohols are often good crystallization solvents, and it is useful to note that alcohols of higher molecular weights tend to have higher boiling points and lower polarities. Consequently, if a compound is too soluble in methanol or ethanol, it is worth trying propanol or butanol. In many cases, a mixed solvent system will be required, and combinations of toluene, or ethyl acetate, with a petroleum ether fraction of similar boiling point can be used. Consult Appendix 1 for boiling points, polarities (dielectric constants), and toxicities of common solvents.
2. Dissolve the compound in the minimum volume of hot solvent. *Remember that most organic solvents are extremely flammable and that many produce very toxic vapors.*

Place the crude compound (where possible, always keep a few seed crystals) and a few boiling chips in a conical flask fitted with a reflux condenser. Then add a small portion of solvent and warm the mixture until a gentle reflux is observed. Add further portions of hot solvent at intervals until all the crude material has dissolved. If you require an antisolvent to aid crystallization, it is then added hot in portions until the compound of interest just starts to precipitate (cloudiness is observed). At this point, a small volume of the good solvent is added to dissolve the initial precipitate, and the hot solution is left to cool. It is important that the crude compound is not contaminated with an insoluble material such as silica or magnesium sulfate, otherwise you may be misled into adding too much solvent during this process.

A conical flask is generally preferred over a round-bottom flask because it is easier to collect the crystals at the end of the process with the former. A jacketed vessel with a bottom runoff valve (Figure 4.15) can also be used and is especially useful for controlled crystallization on larger scales.

3. Filter the hot solution to remove insoluble impurities. This step is often problematic and should not be carried out unless an unacceptable (use your judgment) amount of insoluble material is suspended in the solution. The difficulty here is that the compound tends to crystallize during filtration, so an excess of solvent (~5%) should be added and the apparatus used for filtration should be preheated to just below the boiling point of the solvent. Use a clean sintered funnel of porosity 2 or 3, or a Hirsch or Büchner funnel, and use the minimum suction needed to draw the solution rapidly through the funnel. If the solution is very dark and/or contains small amounts of tarry impurities, allow it to cool for a few moments, add approximately 2% by weight of decolorizing charcoal, heat to reflux for a few minutes, and then remove the charcoal by filtration. Charcoal is very finely divided, so it is essential to put a thin (1 cm) layer of a filter aid such as Celite in the filter funnel to prevent clogging of the sinter. Observe the usual precautions for preventing crystallization in the funnel. Often, very dark or tarry products can also be removed by passing a solution of the crude product through a short (2–3 cm) plug of silica prior to crystallization.

4. Allow the solution to cool and the crystals to form. This step is usually straightforward except when the material is very impure or has a low melting point (<40°C) in which case it sometimes separates as an oil. If an oil is formed, it is best to reheat the solution and then allow it to cool slowly. Try scratching the flask with a glass rod or add a few seed crystals to induce crystallization. If this fails, try adding some more solvent so that precipitation starts at a lower temperature. If nothing at all precipitates from the solution, try scratching the flask with a glass rod, seeding, or cooling the solution in ice water. If all these techniques

fail, stopper the flask and leave in the freezer overnight; patience is sometimes the best policy. Do not be afraid to try alternative solvent systems if you have problems or your crystals are of poor quality.

5. Filter off and dry the crystals. When crystallization is complete, collect the crystals by filtration using an appropriately sized sintered glass funnel. It is very important to wash the crystals carefully. As soon as all the mother liquor has drained through the funnel, remove the suction and pour some *cold* solvent over the crystals, stirring them if necessary, in order to ensure that they are thoroughly washed. Drain the washings under suction, and repeat the procedure once or twice more. After careful washing, allow the crystals to dry briefly in air and then remove the last traces of solvent under vacuum in a vacuum oven, in a drying pistol, or on a vacuum line. Take care to protect your crystals against accidental spillage or contamination. If they are placed in a dish, beaker, or sample vial, cover them with aluminum foil, secure with a wire or an elastic band, and punch a few small holes in the foil. If the crystals are in a flask connected directly to a vacuum line, use a tubing adapter with a tap and put a plug of glass wool in the upper neck of the adapter so that the crystals are not blown about or contaminated with material from the tubing when air or an inert gas is allowed in. If a solvent with a relatively high boiling point such as toluene is used for crystallization, it may be necessary to warm the sample under vacuum for several hours to ensure that all the solvent is removed.

11.2.2   Small-scale crystallization

When small samples (<200 mg) are to be crystallized, it is particularly important to remove traces of insoluble material from the solution and to avoid contamination of the crystals by dust, filter paper, etc. Thus, it is essential to use carefully cleaned glassware and purified solvents and to filter the hot solution. The apparatus shown in Figure 11.1 is convenient for small-scale (5–200 mg) crystallizations. Place the sample in the bulb and wash it with a few drops of solvent. The procedure is as follows:

1. Heat the bulb in an oil bath or a water bath so that the solvent refluxes up to the level of the sintered disk. Add more solvent, in small portions, until the compound is dissolved.
2. Remove the apparatus from the bath, wipe the bulb to remove oil or water, and quickly filter the hot solution into a clean receiver by pressurizing the vessel using a hand bellows or an inert gas line. Filtering under pressure in this way avoids the problem of unwanted crystallization and reduces transfer losses. The hot solution can be filtered into a small conical flask, but it is on a scale of 100 mg or less; a Craig tube (Figure 11.2) gives better recovery because it allows the crystals to be isolated without another filtration.

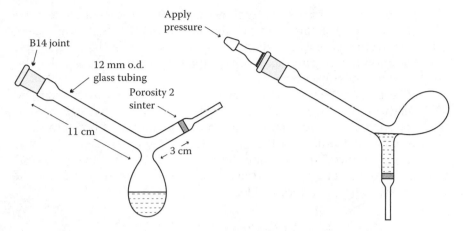

Figure 11.1 Apparatus for small-scale recrystallization.

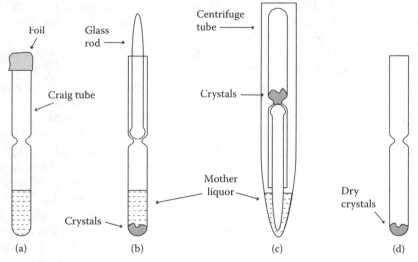

Figure 11.2 Using a Craig tube: (a) filter the hot solution into the tube, (b) insert rod when crystallzation is complete, (c) invert the Craig tube in a centrifuge tube and centrifuge, (d) dry crystals remain in the Craig tube.

3. Filter the hot solution into a suitably sized Craig tube and cover the tube with aluminum foil while crystallization takes place (Figure 11.2a).
4. When crystallization is complete, fit the matching glass rod (a close fit is essential) into the Craig tube and secure it tightly with a rubber band (Figure 11.2b). Place the inverted assembly in a centrifuge tube and centrifuge for a few minutes (remember to use a counterbalancing tube).

5. Centrifugation forces the mother liquor into the centrifuge tube (Figure 11.2c). Remove the Craig tube from the centrifuge tube, remove the glass rod, and then tap the tube so that the crystals fall to the bottom (Figure 11.2d). Cover the tube with a foil, and dry the crystals under vacuum.

6. If necessary, the product can be recrystallized in the same tube by repeating the aforementioned procedure. The crystals can be dissolved in the Craig tube in the minimum volume of hot solvent as usual, and the neck of the tube will act as a condenser. However, care is required and the tube should be no more than one-third full of solvent in order to avoid losses. This process can be carried out several times with minimal losses.

11.2.3 *Crystallization at low temperatures*

For the purification of a low-melting solid, or a thermally unstable liquid, low-temperature crystallization is occasionally useful. However, this technique is attended by two important complications: First, cooling the materials and apparatus below ambient temperature will cause condensation inside the flask. This could be a problem if the compound is sensitive to water. The crystallization flask can be fitted with a drying tube, but it is usually better to carry out the crystallization under an inert atmosphere. The biggest challenge with low-temperature crystallization is keeping the solution cold during the filtration and washing steps. This can be achieved by using apparatus that can be immersed in a cold bath, by using specially built apparatus with integral jackets for holding cooling solutions or, in small-scale work, by carrying out the filtration and washing steps very rapidly. A number of ingenious solutions to these technical problems have been developed, but only a few methods that involve standard organic laboratory glassware are described here.

For medium- to large-scale crystallizations (≥ 1 g), a setup as shown in Figure 11.3 can be used. The compound is dissolved in the minimum volume of solvent at room temperature and is filtered into a two- or three-neck pear-shaped flask. The flask is then fitted with an inert gas inlet and a thermometer adapter containing a filter stick connected to a bubbler. With the filter stick held above the solution, the flask is purged with inert gas and is placed in a cooling bath. It should be cooled slowly by gradually adding a cooling agent to the solvent. When crystallization is complete, the bubbler is disconnected and the filter stick is connected to an appropriately sized receiver using chemically inert tubing (PTFE). The filter stick is then lowered into the solution and the mother liquor is forced into the receiver by pressure exerted by the inert gas. The thermometer adapter allows sufficient freedom of movement of the filter stick to pack the crystals to the bottom of the flask and drain off the mother liquor thoroughly. The crystals can then be washed by releasing the pressure and adding small amounts

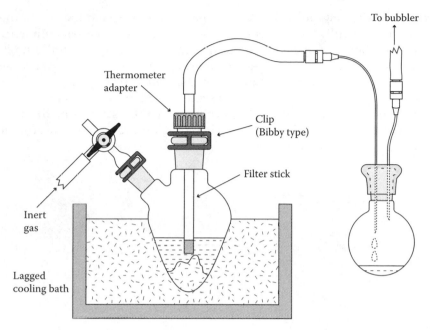

Figure 11.3 Recrystallization under an inert atmosphere.

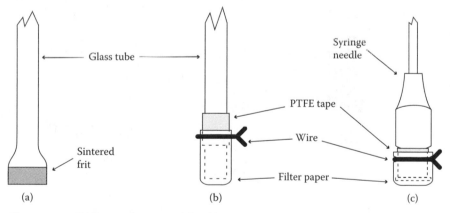

Figure 11.4 Different designs of the filter stick: (a) glass tube with sintered glass frit, (b) glass tube with filter paper, and (c) syringe needle with filter paper.

of precooled solvent via the three-way tap using a cannula. The washings can then be removed using the filter stick as before. The cooling bath can be removed and the crystals isolated and dried in the usual way, or the low-temperature crystallization can be repeated in the same flask.

The sintered disk (Figure 11.4a) of the filter stick should be of porosity 3 or larger in order to avoid blockage. A convenient alternative to using a filter stick is to use an ordinary glass rod with filter paper wrapped tightly round the end (Figure 11.4b). Wrap some PTFE tape round the end of the tube; then carefully fold a filter paper over the end and secure it with wire. The PTFE tape will give a better seal because the wire will sink into it. Another alternative, for small-scale work, is to use a long syringe needle (Figure 11.4c). Again, the end of the needle should be covered with filter paper, wrapped with PTFE tape, and secured with wire. The sharp end of the needle can be inserted into the receiving flask, thus obviating the need for connecting tubing (Figure 11.5). These devices are also very useful for filtering reaction mixtures under an inert atmosphere.

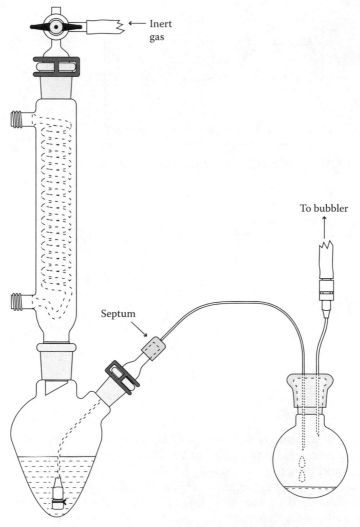

Figure 11.5 Using a filter stick made from a syringe needle.

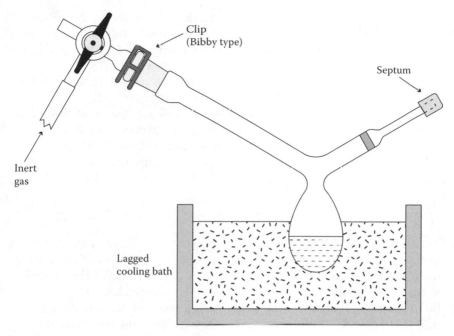

Figure 11.6 Low-temperature recrystallization.

Small-scale low-temperature crystallizations can be carried out in the same device recommended for ordinary small-scale work (Figure 11.6). First of all, dissolve the material in an appropriate solvent in the bulb. Then purge the apparatus with an inert gas and seal the outlet with a small septum. Immerse the flask in a cooling bath; when crystallization is complete, remove the septum and filter the suspension rapidly under inert gas pressure. On a small scale, the solution will not warm up significantly during the short time required to filter off the crystals. The crystals can be washed with precooled solvent.

11.2.4   Crystallization of air-sensitive compounds

Clearly, the crystallization of air- and moisture-sensitive compounds must be carried out under an inert atmosphere. The trickiest step in this process is often the introduction of the crude solid to be crystallized into the crystallization flask. This is usually best achieved by transferring the material in solution. Alternatively, it may be possible to transfer a solid very quickly in air or using a blanket of inert gas under an inverted funnel (Figure 6.20). For very sensitive compounds, transfer of solids needs to be carried out in a glove bag or a glove box.

Once the solid is placed in the flask, the methods described in Section 11.2.3 for low-temperature crystallizations under an inert atmosphere can be used. The apparatus shown in Figures 11.3, 11.5, and 11.6 can be used without the cooling bath, unless it is required.

One particular problem that arises with compounds that hydrolyze on contact with moisture is that small amounts of decomposition products can block filters. Thus, if a sintered disk is used, it must be scrupulously dry to prevent the occurrence of hydrolysis in the pores of the filter disk. In fact, filter sticks made using filter paper are often better than sintered disks. It is good practice to keep at least one backup dry filter on hand in case the first one becomes blocked.

11.3   Distillation

Distillation is the most useful method for purifying liquids, and it is used routinely for purifying solvents and reagents. Using appropriate apparatus, and with some care, it is possible to separate liquids whose boiling points differ by less than 5°C. We assume that the reader is familiar with the fundamentals of the theory and practice of distillation, but it is appropriate to begin discussing this topic by reiterating some basic safety rules:

1. Never heat a closed system.
2. Remember that most organic liquids are extremely flammable, so great care must be taken to ensure that vapors do not come in contact with flames, sources of sparks (e.g., electrical motors), or very hot surfaces (hot plates).
3. Never allow a distillation pot to boil dry. The residues may ignite or explode with great violence.
4. Beware of the possibility that ethers and hydrocarbons may be contaminated with highly explosive peroxides. Be particularly careful when distilling compounds prepared by peroxide and peracid oxidations, and always take precautions to destroy peroxide residues prior to distillation.
5. Carry out a safety audit on the compound you plan to distill to check that it is not thermally unstable. Some types of compounds, for example, azides, should *never* be distilled.

11.3.1   Simple distillation

The conventional apparatus for simple distillation is shown in Figure 11.7. It is useful only for distilling compounds from nonvolatile residues or for separating liquids whose boiling points differ by at least 50°C. Moisture can be excluded by attaching a drying tube to the vent or by connecting

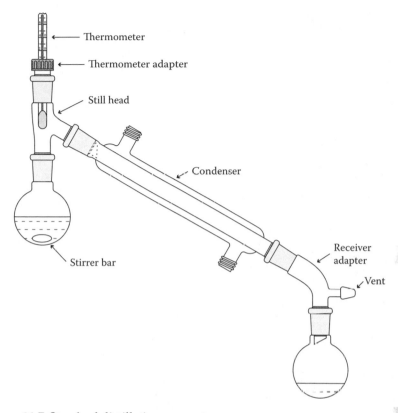

Thermometer

Thermometer adapter

Still head

Condenser

Stirrer bar

Receiver adapter

Vent

Figure 11.7 Standard distillation apparatus.

the vent to an inert gas line. It is essential to add some boiling chips, or stir the liquid, in order to prevent bumping, especially if a finely divided solid, such as a drying agent, is present.

The flask should be heated in a water bath, or an oil bath, and *not* with a Bunsen burner or a heating mantle. The temperature of the bath should be increased slowly until distillation begins, and then it should be adjusted to give a steady rate of distillation. If the boiling point is high (>150°C), the still head may need to be lagged with glass wool, and an air condenser should be used rather than a water condenser.

For most purposes, it is more efficient and convenient to use a compact (short-path) distillation apparatus such as the one shown in Figure 11.8. Such apparatus can be bought or be constructed in two or three convenient sizes to fit common ground glass joints. The main advantage of such devices is that very little material is lost on the sides of the apparatus and joints. Therefore, they are particularly appropriate, if not essential, for small-scale work.

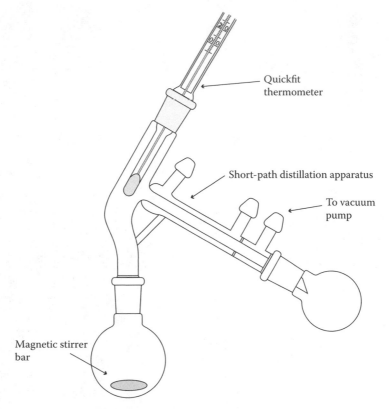

Figure 11.8 One-piece distillation apparatus.

11.3.2   *Distillation under an inert atmosphere*

If distillation is used to dry a reagent, the process must be carried out under an inert atmosphere. The following procedure can be used:

1. Dry all the glass apparatus in an oven, or with a heat gun under vacuum, and purge with inert gas while cooling. This is most easily accomplished by connecting the apparatus to a double manifold/bubbler system (see Chapter 9, Section 9.2.3). Although Quickfit distillation assemblies can be used, we recommend the one-piece apparatus shown in Figure 11.8.
2. When the glassware has cooled, increase the inert gas flow, quickly disconnect the distillation flask, add any drying agent required, a few antibumping granules (or a stirrer bar) and the liquid to be distilled, then reassemble the apparatus.

3. Heat the distillation flask *in an oil bath* (do not carry out distillations using a heating mantle), and collect the liquid that distills at the required temperature.
4. When the distillation is complete, remove the collecting flask and seal it quickly with a septum. Most reagents can simply be poured into a reagent bottle before sealing provided you are quick. However, if the reagent is particularly sensitive to air or moisture, a cannulation technique should be used to transfer it (see Chapter 6, Section 6.4.2). Whatever the type of container used for storage, it is always preferred that the container is full or nearly full.

11.3.3 Fractional distillation

Separation of liquids whose boiling points differ in the range 5°C–50°C requires the use of a fractionating column, which gives better contact between the vapor and liquid phases in a distillation column. One of the most common types of fractionating columns, a Vigreux column, is shown in Figure 11.9a. A 50 cm long vacuum-jacketed Vigreux should allow reasonable separation of compounds whose boiling points differ by 30°C–40°C.

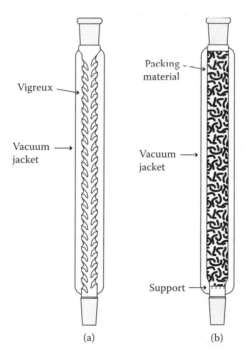

Vigreux

Vacuum jacket

Packing material

Vacuum jacket

Support

(a)   (b)

Figure 11.9 Fractionating columns: (a) Vigreux column and (b) packed column.

One-piece assemblies incorporating short Vigreux columns (Figure 11.10) are somewhat less efficient but very convenient for routine distillations and give good results at reduced pressure.

Efficient separation of compounds with a boiling point difference of 10°C–30°C can be achieved using a long glass tube packed with glass rings or helices or, for high efficiency, wire mesh rings (Figure 11.9b). The key to getting good results from a fractional distillation is to raise the temperature gradually and collect the distillate very slowly.

Two, more sophisticated commercially available designs are Spaltrohr columns, which consist of concentric grooved tubes, and spinning band columns, which have a rapidly spinning spiral band fitted inside the column. Both these systems have low hold ups and give very high efficiency, so they can be used for the separation of small volumes (as little as 1 cm³) and for separating compounds (with differences as little as 3°C). Consult the manufacturers' manuals for operating instructions for these devices.

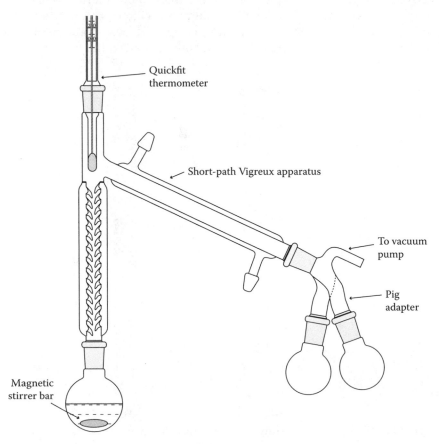

Figure 11.10 One-piece Vigreux distillation apparatus.

11.3.4   *Distillation under reduced pressure*

Many compounds decompose when heated to their boiling points, so they cannot be distilled at atmospheric pressure. In this case, it may be possible to avoid thermal decomposition by carrying out the distillation at a reduced pressure. The reduction in boiling point will depend on the reduction in pressure and can be estimated using a temperature/pressure nomograph (Figure 11.11).

To find the *approximate* boiling point of a compound at any pressure, simply place a ruler on the central line of the nomograph at the atmospheric boiling point of the compound, pivot it to line up with the appropriate pressure on the right-hand line, and read the predicted boiling point from the left-hand line. You can also use the nomograph to find the boiling point at any pressure if you know the boiling point at some other pressure by using the known data to estimate the boiling point at atmospheric pressure. Note that although pressure is usually measured in millimeters of mercury (mmHg), it is often quoted in different units, especially Torr. Happily, 1 Torr = 1 mmHg. Boiling points measured at reduced pressure may be expressed in several ways, for example, 57°C/ 25 mmHg or B.p.$_{25}$ 57°C. As a very rough guide, a water pump (~15 mmHg) will give a 125°C reduction in boiling point and an oil pump (~0.1 mmHg) will give a reduction of 200°C–250°C.

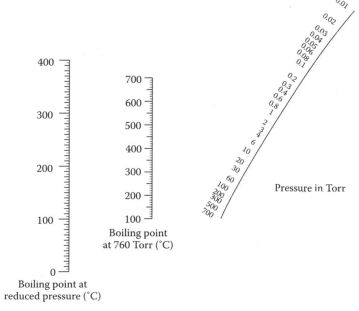

Figure 11.11 A temperature/pressure nomograph.

A typical vacuum distillation apparatus is shown in Figure 11.10. Its chief difference from the simple distillation apparatus is in the design of the receiver adapter. This receiver adapter must allow several fractions to be collected without needing to break the vacuum. The simplest design is the "pig" type shown in Figure 11.12a. When using a pig, remember to grease the joint lightly, around the outer edge only, so that the receiver can be rotated while under vacuum. Also, fix the receivers securely using clips.

A procedure for carrying out a distillation under reduced pressure is as follows:

1. Place the sample in the distillation flask (no more than two-thirds full), and add a stirrer bar. *Note:* Boiling chips are not effective under reduced pressure, and so an alternative must be used. A very narrow capillary attached to an inert gas line, which allows a slow stream of nitrogen bubbles to pass through the solution, is effective, but brisk stirring using a magnetic follower is usually more convenient.
2. Assemble the (oven-dried) apparatus, putting a small quantity of high vacuum grease on the outer edge of each joint. Ensure that the receiver adapter and the collection flasks are secured using clips,

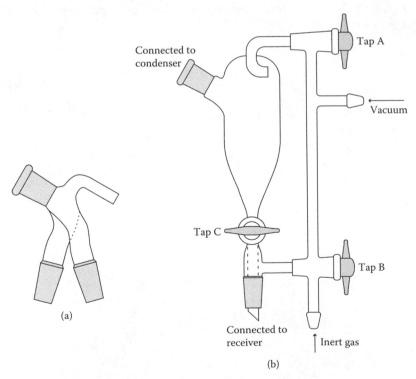

Figure 11.12 Figure showing (a) pig and (b) Perkin triangle apparatus.

and connect the assembly to a vacuum pump. One convenient method of doing this is to connect the assembly to a vacuum/inert gas double manifold (see Chapter 9, Section 9.2.3). The pump must be protected with a cold-finger trap, and the line should incorporate a vacuum gauge for monitoring the pressure (see Chapter 8, Section 8.4).

3. Stir the liquid rapidly and *carefully* open the apparatus to the vacuum. Some bumping and frothing may occur as air and volatile components are evacuated. If necessary, adjust the pressure to the required value by allowing an inert gas into the system via a needle valve.

4. Heat the flask slowly to drive off any volatile impurities and then distill the product. Monitor the still head temperature and collect a forerun and then a main fraction, which should distill at a fairly constant temperature. If fractionation is required, you may need to collect several fractions and it will be important to distill the mixture slowly and steadily.

5. Halt the distillation when the level of liquid in the pot is running low by removing the heating bath.

6. Isolate the apparatus from the vacuum and carefully fill it with the inert gas. If you are using a double manifold, this can be done by simply turning the vacuum/inert gas tap. The flask containing the distillate will then be under a dry, inert atmosphere and can be quickly removed and fitted with a tightly fitting septum.

7. Switch off the pump and clean the cold trap.

A Perkin triangle (Figure 11.12b) is a more convenient device for collecting fractions from vacuum distillations on a larger scale. It is operated as follows:

1. Attach the condenser from the distillation apparatus and a receiver flask to the Perkin triangle. Then open both the distillation apparatus and the receiver flask to the vacuum using taps A and B.

2. When distillation begins, open tap C to collect the forerun in the receiver flask.

3. Close tap C so that the next fraction will be collected in the bulb, while the receiver is being replaced. Use tap B to allow inert gas/air into the receiver flask and then replace the flask with a new one.

4. Close tap A to briefly isolate the still; then evacuate the receiver by opening tap B to the vacuum. When the pressure has steadied, open tap A again. Then tap C can be opened to allow the distillate to drain into the new receiver.

5. Continue to collect fractions by repeating steps 3 and 4 as necessary.

11.3.5 Small-scale distillation

The chief difficulty with small-scale distillations is that a significant propor-tion of the sample may be lost in "wetting" the surface of the column and the condenser. This problem is reduced by using very compact one-piece short-path designs, but this reduces the fractionating efficiency of the columns, leading to less effective separation. Different apparatus designs giving dif-ferent trade-offs between recovery and efficiency are available. A typical example, featuring a short Vigreux and a rotary fraction collector, is shown in Figure 11.13. Spinning band and Spaltrohr columns combine high effi-ciency with low hold up, so they can provide an effective, if expensive, solu-tion to the problem of fractionation of small volumes (as little as 1 cm^3).

Another popular method for distilling small quantities is to use a Kugelrohr apparatus (Figure 11.14). The key features of this system are a horizontal glass oven with an iris closure, that heats the flasks efficiently; a series of bulbs that can be moved horizontally in and out of the oven permitting short-path distillation; and a motor to rotate the bulbs. The Kugelrohr apparatus can be used to distill small quantities (<100 mg) and operated at high vacuums. A simple distillation is carried out as follows:

1. Place the sample in the end bulb using a Pasteur pipette. If neces-sary, wash the sample into the bulb and then evaporate the solvent on a rotary evaporator.
2. Add two more bulbs and connect them to the motor assembly using a straight length of glass tubing.

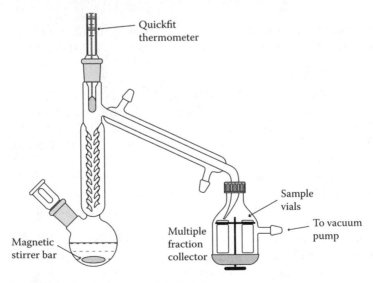

Figure 11.13 Small-scale fractional distillation.

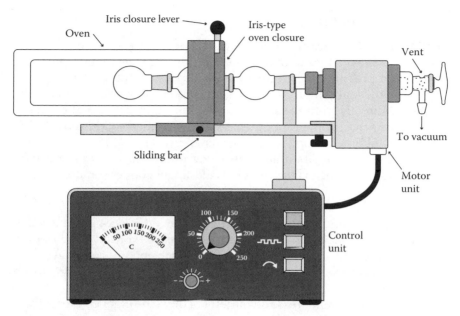

Figure 11.14 Kugelrohr apparatus.

3. Slide the end bulb containing the sample into the oven and gently close the iris seal around the connecting joint.
4. Set the bulbs to rotate (to prevent bumping and to speed up the distillation).
5. Apply the vacuum gradually to prevent frothing and raise the oven temperature gradually until distillation begins (usually indicated by a mist appearing in the first bulb). Adjust the temperature so that distillation continues at a steady rate. In some cases, you may need to cool the collection bulb in order to efficiently condense the product.
6. When distillation is complete, allow the flasks to cool under an inert atmosphere; then remove the bulbs from the apparatus and recover the distillate. If you do not want to wash the collection bulb with the solvent, simply clamp it in a vertical position and allow its contents to drain into a flask or sample vial.

Fractionation of two compounds with a 20°C–30°C boiling point difference can be achieved by inserting two of the three bulbs into the oven, distilling the most volatile component into the third bulb, and then withdrawing the next bulb from the oven and distilling the higher boiling fraction into that bulb.

11.4   Sublimation

Sublimation is an excellent method for purifying relatively volatile organic solids on scales ranging from a few milligrams to tens of grams. At reduced pressure, many compounds, especially those of low polarity, have sufficiently high vapor pressures to be sublimed, that is, to be converted directly from the solid phase to the vapor phase without melting. Condensation of the vapor then gives the purified solid product. Two designs of sublimation apparatus are shown in Figure 11.15.

The larger sublimation apparatus (Figure 11.15a) consists of a tube with a side arm, which is fitted with a cold-finger condenser; it is used as follows:

1. Place the crude solid in the bottom of the outer vessel. If it is waxy or oily, wash it into the tube with a *small amount* of solvent, cover the side arm with a septum, and remove the solvent on a rotary evaporator. If it is a dry solid, grind it into a fine powder before adding to the vessel.
2. Put some vacuum grease on the joint of the cold-finger condenser and fit it into the sublimation apparatus. There should be a gap of approximately 1 cm between the solid and the bottom of the condenser.

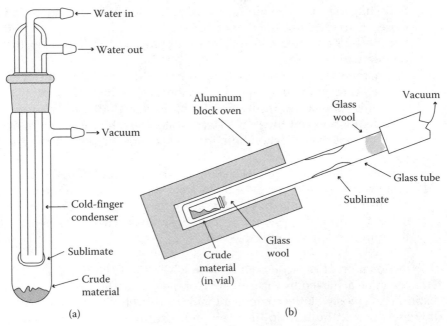

Figure 11.15 Sublimation apparatus: (a) cold-finger sublimation and (b) small-scale sublimation oven.

3. Evacuate the apparatus slowly to prevent any spattering of the solid. Turn on the condenser water, and slowly heat the base of the sublimation apparatus.
4. A fine mist of sublimed material on the condenser indicates that sublimation is beginning and, subsequently, the temperature should be held fairly constant until the process is complete.
5. When sublimation is complete, the product may be clinging precariously to the cold finger, so proceed with great care. Turn off the water, *carefully* allow air/inert gas into the sublimation apparatus, and *very carefully* remove the cold finger. Scrape off the product with a microspatula.

The simple glass tube sublimator shown in Figure 11.15b functions in a similar way as the sublimator shown in Figure 11.15a except that the water condenser is omitted in the former. Place the crude sample in a small sample vial, put a plug of glass wool in the neck, and drop the vial into the tube. Put a plug of glass wool in the neck of the tube and attach it to a vacuum line. Evacuate carefully and heat the base of the tube gently in an oil bath or a custom-designed metal block as shown. Proceed as before and isolate the sublimed product by cutting the tube and scraping out the solid.

11.5   Flash chromatography

Chromatographic techniques for analysis and purification of reaction products are probably the most universally important of all the skills required by an organic chemist.

Before attempting any of the chromatographic separation techniques described in the rest of this chapter, you need to be confident about your ability to run analytical TLC (Chapter 9, Section 9.3.1) as the two skills are very closely interlinked. For routine separation and purification of reaction products, some form of column chromatography is normally employed. Traditionally, long columns were filled with silica and a good head of solvent was used so that a flow was achieved by force of gravity. This method of chromatography is very slow and the slow elution rate leads to band dispersion. This reduces resolution and often leads to a large number of fractions that contain a mixture of compounds. The technique has therefore become obsolete and has been superseded by flash chromatography.

Flash chromatography was introduced by Still, Khan, and Mitra in 1978.[1] It has cut down the time taken for routine purification of reaction mixtures considerably and, perhaps more importantly, has provided the chemist with a fast and simple technique for separating isomeric materials having similar polarities. It has become the standard method of purification in many synthetic laboratories, with a resultant reduction in the time taken to achieve many synthetic goals. The key to the effectiveness of flash chromatography is the comparatively fine silica powder that is used with a relatively narrow particle range (typically 40–60 μm). This silica powder gives better surface contact, and therefore more effective adsorption, than that previously used for gravity columns. Using pressure to drive the solvent through the column relatively quickly increases the resolution by cutting down band dispersion and also reduces the time required.

There are some practical difficulties associated with the originally published method for running flash columns, and modifications have therefore been introduced over the years.[1] These make the technique easier to carry out and somewhat safer. Sections 11.5.1 through 11.5.3 will lead you through the basic steps for running a version of flash chromatography that we find to be both convenient and effective. Flash chromatography is a very powerful and fast technique for separating organic compounds. Like all chromatographic techniques, you can become skilled in it only through experience, and the instructions in the following subsections should be considered as a starting point. Every separation is different, so do not be despondent if you have some failures in the beginning. With practice, you should be able to acquire the skill and intuition required to get a good separation every time and even in the most difficult circumstances.

11.5.1   Equipment required for flash chromatography

The setup originally described by Still requires the use of long columns to accommodate a reasonable supply of solvent, but these are difficult to load and need to be dismantled for adding extra solvent during a run. We therefore recommend the use of shorter columns plus a reservoir. It is important to have a familiar set of columns on hand, and we recommend that you have a set of about five columns ranging in diameter from about 5 to 50 mm. A very convenient length for all columns is 25 cm except for columns with very narrow bores, which can be shorter (Figure 11.16). It is convenient to make B24 joints the standard for all columns and reservoirs so that they are interchangeable. A glassblower can make reservoirs from a round-bottom flask and a male joint. We use a 500 cm³ reservoir more commonly than any other, but it is also useful to have 100 cm³, 250 cm³, and 1 L sizes available. Although equipment fitted with ordinary Quickfit joints, secured by rubber bands, can be used, equipment fitted with Rodaviss joints is recommended. A standard male joint is secured into a Rodaviss socket by a screw collar and a rubber O-ring. This is a far superior method for connecting glassware for flash chromatography in terms of both convenience and safety.

In our experience, it is difficult to control the column pressure using a gas cylinder in conjunction with a Rotaflow valve as suggested in the

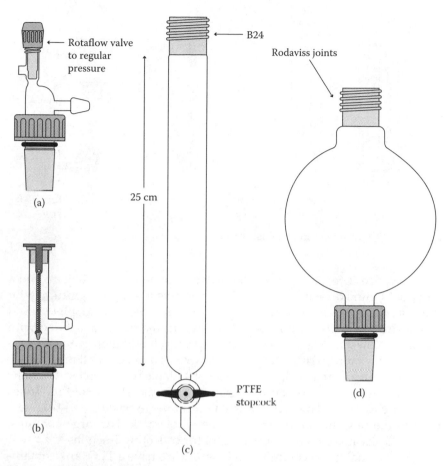

Figure 11.16 Flash chromatography column and solvent reservoir: (a) flash adapter, (b) flash valve, (c) flash column, and (d) reservoir.

original paper and can be dangerous if too much pressure is inadvertently applied, and so is best to avoid using a gas cylinder and Rotaflow valve.[1] There are various alternative ways in which pressure can be applied to the column in a safer way. One method is to use a small (fish tank–type) low-pressure diaphragm pump. If an appropriate model is used, it will give a good flow rate without causing too much pressure to build up. Another safe and simple method for pressurizing columns is to use rubber bellows. The only problem with this technique is that it is difficult to keep the pressure constant, particularly with large columns.

If compressed gas is used to pressurize flash columns, we recommend that some sort of pressure-release valve is incorporated to prevent high pressures from building up. A simple pressure-release valve, such as the "flash valve" shown in Figure 11.17, can be constructed easily and is

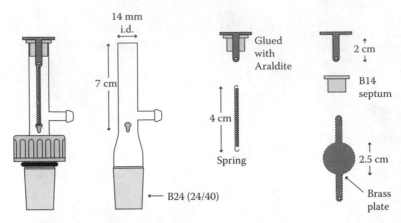

Figure 11.17 Construction of a flash valve.

very effective. The glass part of this valve is made from a B24 joint and a piece of 14 mm i.d. heavy wall tubing. The valve is simply a septum, with the skirt cut off, glued to a shaped metal (brass) plate. The adapter should be constructed with the dimensions shown in the figure and fitted with standard 4 cm stainless steel laboratory springs. It will then provide a constant safe pressure, which will give a flow rate suitable for all purposes.

For relatively small columns, the most suitable way to collect fractions is in a rack of test tubes, and the best racks are those that hold the tubes very close together. This will allow column flow to be changed from one tube to the next simply by moving the whole rack. For large columns, conical flasks are often more convenient for collecting fractions.

For the analysis of column fractions, always have a TLC tank containing an appropriate solvent and several TLC plates cut to about 4 cm wide close at hand.

11.5.2   Procedure for running a flash column

Safety note: Silica dust is very toxic if inhaled; therefore, you should take precautions to avoid breathing it in and always handle silica in a fumehood. Large volumes of solvents are also used for chromatography, and you should take precautions to avoid breathing in their vapors or exposing them to sparks.

Before starting a flash column, you need to determine the amount of silica gel required and the solvent system to be used. This is done as follows:

1. First, run a series of TLCs to find a solvent system that will give a good separation of components of the mixture under study. For the majority of organic molecules, it is usually best to start with

ethyl acetate–petroleum ether mixtures. If several components of interest are running close together, an R_f value of 0.2–0.3 at the midpoint between them will indicate a satisfactory solvent system. If they are well separated, a solvent that puts the lower spot at an R_f of 0.2–0.3 will usually work. If you know which spot you are most interested in, try to bias your judgment toward this spot. There are often irrelevant impurities that are either very polar or very nonpolar and they can be largely ignored.

2. You next need to decide on the amount of silica to be used. Always try to use as little silica as possible since it is quite expensive. Where the component you require is well separated from other components, a ratio of approximately 20:1 (silica:sample) should be sufficient. For more difficult separations, ratios up to 100:1 can be used. With this ratio, a skilled researcher should be able to separate spots that are overlapping on TLC. Figure 11.18 provides a rough guide of silica:sample ratios that should work for different situations. An important fact to remember for all chromatographic separations is that the more spots there are in a mixture the greater the ratio of

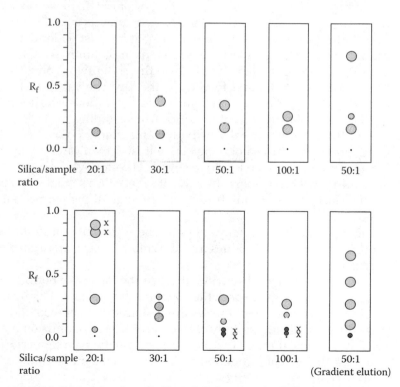

Figure 11.18 Determination of silica:sample ratios for flash chromatography. It is noted that spots marked "x" are not required products.

silica to each individual component of the mixture. Thus, you normally require *less*, not more, silica to separate a three-spot mixture than a two-spot mixture with similar separations.

The above two steps are the key to a successful separation using flash column chromatography. It is in this part of the procedure that the greatest skill and judgment is required and careful attention should be paid to trial TLCs. With experience, you are likely to develop an instinct for the appropriate quantity of silica and the solvent system.

Next, you need to select the appropriately sized column and prepare it. This is done as follows:

1. Measure out the required quantity of silica in a conical flask. You can do this by weighing the silica, but if you are using a graduated conical flask it is often more convenient to estimate the weight by volume of silica (dry silica has a density of ~0.5 g/cm^3).
2. Select a column that will fill to a depth of about 18 cm with the amount of silica being used (it is useful to label columns once you know how much they hold). Plug the bottom of the column with fresh (clean) cotton wool. The easiest way to do this is to roll a piece of cotton wool between your fingers so that it is just wider than the column outlet. Then connect the bottom of the column to a low vacuum line keeping the tap closed; drop the cotton wool plug to the bottom of the column and then open the tap. The vacuum should suck the plug into place. Make sure that the cotton wool forms a good enough plug to prevent the silica from escaping.
3. Mount the column vertically using a clamp stand, pour in about 8 cm depth of the chromatography solvent, and then carefully sprinkle a layer of fine sand (~1 mm) to cover the cotton wool plug.
4. Add sufficient chromatography solvent to the silica so that it forms a mobile slurry. Swirl the mixture to ensure that all the trapped air is removed.
5. Using a powder funnel, *carefully* pour the silica slurry into the column. Wash any silica residues into the column using more chromatography solvent.
6. When all the silica has been added, open the tap at the bottom and pressurize the column to pack the silica. It is important to ensure that no air is trapped in the column as this will lead to poor resolution. If you see air bubbles in the body of the column, let a column volume of solvent percolate through the silica under gravity; then repressurize and flush the solvent through until no air remains in the column. *Be careful not to allow the solvent to drop below the level of the top of the silica at any point.* Leave a good head of solvent above the silica and sprinkle enough sand to cover the surface of the silica evenly (~1 mm).

Then force the excess solvent through the column until there is just a small layer (~1 mm) left above the sand. It is essential that the top of the silica is completely flat, so take extreme care when carrying out this part of the operation.

Once prepared, the column can be left for a short time; but the solvent level must not be allowed to drop below the level of the sand. Once you load the sample, the column should not be left to stand and the remaining steps should be carried out as swiftly as possible. For best results, the solvent flow should be continued without interruption, especially during the early stages of elution. Leaving the column standing with the sample loaded will lead to band dispersion and loss of resolution. So, before starting the remainder of the procedure make sure you have everything you need at hand, including plenty of solvent and TLC plates.

The next stage is to load the sample onto the column, but before doing this remember to keep a TLC sample of the original mixture. The sample is loaded as follows:

1. Dissolve the sample in the minimum amount of solvent. It is best to dissolve the sample in the same solvent system that is going to be used for the column. If this is not possible, and you are using a solvent mixture, try adding a few extra drops of the more polar solvent. If this fails to dissolve, there are two options: You could use a different solvent (e.g., dichloromethane) to dissolve the same. This will usually work if you are using a relatively polar solvent system such as 50% ethyl acetate 50% petroleum ether for the column. However, if the solvent system is less polar, even small quantities of dichloromethane may cause elution problems as it will drag more polar components through the column. To avoid this, you can dry load the column. This is done by mixing a small amount of silica (enough to form a 3 mm layer at the top of the column) to a solution of the sample in solvent. The mixture should then be evaporated to dryness using a rotary evaporator.

2. Load the solution (or dry sample-impregnated silica) onto the top of the column *very carefully* using a Pasteur pipette (Figure 11.19a). It is essential that you do not disturb the flat surface of silica at the top of the column. If you are adding a solution, it is best to run it down the walls of the column just above the sand. Move the pipette in a circular motion around the internal glass walls of the column to ensure an even distribution of sample on the top of the column. The same approach can be used to add dry sample-impregnated silica, but try to ensure an even layer of silica by carefully adding some to the center of the column as well. Once the sample has been added, open the tap at the bottom of the column and allow the solvent level to drop to

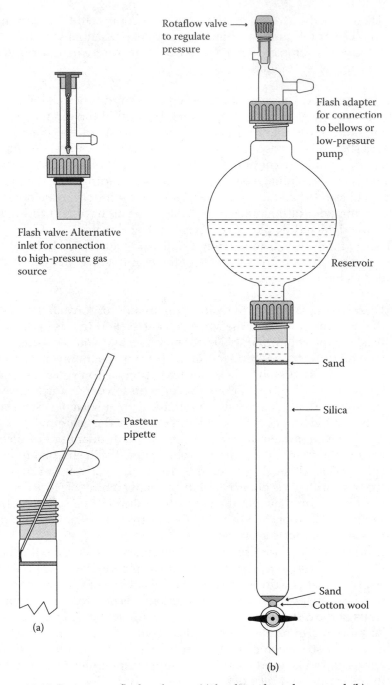

Rotaflow valve →
to regulate
pressure

Flash adapter
for connection
to bellows or
low-pressure
pump

Reservoir

Flash valve: Alternative
inlet for connection
to high-pressure gas
source

Sand

Silica

Pasteur
pipette

Sand
Cotton wool

(a)

(b)

Figure 11.19 Running a flash column: (a) loading the column and (b) complete
column setup.

the top of the sand (or silica). Close the tap; then using the minimum quantity of solvent rinse off any remaining sample from the flask, adding it to the top of the column in the same way. Make sure that any sample that may have stuck to the sides of the column above the sand is washed down onto the top of the column. Finally, allow the solvent to drain flush with the top of the column. The sample is now loaded.

The next stage is to add the solvent to the column:

1. Add the solvent to the top of the column *very carefully*, again using a Pasteur pipette to drip the solvent around the walls of the column just above the sand (Figure 11.19a). It is essential that the surface of the column is not disturbed at any stage of this operation.
2. Once the solvent has reached the joint at the top of the column, attach a solvent reservoir and secure it using a Rodaviss collar or elastic bands (Bibby clips are not strong enough). Then carefully fill the reservoir with the solvent poured from a measuring cylinder.

You are now ready to run the column:

1. Connect the flash adapter (or rubber bellows) to the top of the reservoir and secure it using springs or elastic bands. At this point, your setup should look like the diagram shown in Figure 11.19b.
2. Apply sufficient pressure so as to give a fast solvent flow rate and collect fractions continuously. A pressure of 7 psi is recommended in Still's original paper.[1] If you have a pressure regulator set it to this value, but if not just aim to maintain a fast flow rate through the column (the solvent should run rather than drip). A slow flow rate causes reduced resolution, *not* improved separation. At all stages of column chromatography, be careful not to let the column run dry.
3. Collect your fractions in test tubes or small conical flasks. The ideal size of the fractions collected will depend on the amount of silica used in the column. As a rough guide, they should be (in cubic centimeters) about half the weight of silica used (e.g., if 30 g of silica is used, you should collect 15 cm^3 fractions). It is a myth that you get less mixed fractions by collecting smaller fractions. All that happens is that the mixture is collected in more tubes, requiring more TLCs to be run and thereby extra work for the same result. You may feel safer collecting relatively small fractions at first, but as you become more experienced you should start to see the advantages of collecting larger fractions and thus considerably lowering the amount of time you spend on chromatography.

4. As the column is running, it is important to continually rinse the end of the column (where the solvent comes out) with a small amount of solvent each time you collect a fraction. This is because as the solvent pours out of the column, a small amount of the eluent spreads up the outside of the glassware due to surface tension. This evaporates and leaves a residue of nonvolatile material. When a fraction is eluting this "spreading and evaporation" can leave a significant amount of residue on the tip of the column (if the fraction contains a solid, you will often be able to see a solid deposit on the outside of the glassware). If you do not wash this off into the fraction that is being collected at the time, it will contaminate later fractions.

5. Analyze the column fractions by running TLCs while the column is running. It is best to use TLC plates that are large enough to accommodate five spots and to TLC every other fraction. In this way, you will need to run one plate for every 10 fractions collected. If the column is running correctly, this will give you enough time to spot the second plate while the first one is running. You should also have time to apply a spot of the previous fraction to a TLC plate while the present fraction is being collected. Getting the timing of this right will take a little practice, but once you acquire this skill you should be able to easily complete a column in less than an hour. Figure 11.20 shows an example of what should be possible from a well-run column.

6. Finally, use the TLCs to identify the fractions to combine. The TLC analysis of the fractions will tell you when all the compounds you are interested in have eluted. They will also identify which tubes have which components. If you have run TLCs on every other fraction (as suggested in step 5 above), it may be necessary to run TLCs of some of the intermediate fractions at this stage. Once you are sure what is in each fraction, they should be combined as appropriate (keeping fractions that contain mixtures separate from those containing pure materials). The purified components can then be recovered by removing the solvent from the combined fractions using a rotary evaporator. Any remaining traces of the solvent can then be removed under high vacuum.

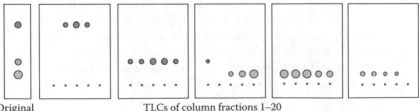

Original
mixture   TLCs of column fractions 1–20

Figure 11.20 The TLCs of an ideal set of fractions from a successful column.

Usually, when the components of a mixture run close together on TLC, a single solvent system that gives the spots an R_f of 0.2–0.3 will be effective. However, when the R_f values of the spots are a long way apart, increasing the solvent polarity as the column is running can save time and solvent. This change in solvent composition as the column is running is referred to as gradient elution. You should be confident with single solvent chromatography before you attempt gradient elution as the latter requires quite a bit of experience to be successful. The procedure for doing gradient elution is as follows:

1. Start running the column in a solvent that will give the highest running spot an R_f of 0.2–0.3.
2. When TLC analysis indicates that this component is almost completely off, change the solvent polarity to that which gives the second spot an R_f of 0.3. Be careful at this point: If the second solvent system is substantially different, you may need to change the solvent polarity in steps. This is because mixing of significantly different solvent systems can be exothermic and may lead to "cracking" of the silica in the column. This will damage the efficiency of the column, leading to poor separation.
3. Continue this process until all the spots that you require are off.

11.5.3 Recycling silica for flash chromatography

The silica used in flash chromatography is quite expensive, but if you can buy it in bulk (25 kg at a time) the cost may be reduced by nearly half. Another way to save money is to recycle the silica using the following procedure:

Remember, silica dust is very toxic if inhaled; therefore, you should take precautions to avoid breathing it in and always handle the silica in a fumehood.

1. Wash the silica. Suspend 1 kg of silica in 1.5 L of acetone, stir, filter using a Buchner funnel, and wash with an additional 1 L of acetone followed by 2 L of deionized water.
2. Place the washed silica in a large crucible and dry at 120°C–140°C overnight. Then heat it in a muffle furnace at 500°C–600°C for 4 hours.
3. After cooling, you need to remove course particles. Place the silica in a bucket containing about 5 L of deionized water. Mix the suspension thoroughly, avoiding vortexing, and allow the mixture to settle for 30 seconds. Then *carefully* pour the top 50% of the mixture into a second bucket. Make the first bucket up to 5 L, stir again, allow it to settle for 30 seconds, and again decant off the first 50% into the second bucket. Repeat this procedure two more times and then discard the remaining contents of the first bucket.

4. The second bucket should now contain about 10 L of a silica/water mixture. You now need to remove fine particles. Stir the mixture thoroughly and allow it to settle for 4 minutes. Decant off the first 50% *carefully* and discard it. Stir again, allow it to settle for 4 minutes, decant off the first 50%, and discard again. Repeat this procedure two more times after making up to 5 L each time.
5. Finally, the silica needs to be reactivated. Filter the silica using a Buchner funnel; then dry it in an oven at 120°C–130°C overnight. Recovery should be 650–800 g.

11.6   Dry-column flash chromatography

This technique was developed by Harwood[2] and can be used as an alternative to flash chromatography. The apparatus consists of a parallel-sided vacuum filter funnel incorporating a porosity 3 sinter and a flask (Figure 11.21).

To run a dry column, you need to do the following:

1. Prepare the column by first filling the funnel to the lip with TLC-grade silica; then tap gently while applying suction from a water aspirator. Press the silica down starting at the circumference and working toward the center. Continue until a level and firm bed is obtained and there is enough head space for sample and solvent addition. The approximate funnel sizes compared to the quantity of the sample to be applied are given in Table 11.1.

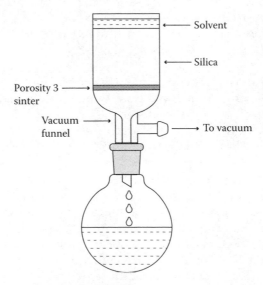

Figure 11.21 Dry-column chromatography.

Table 11.1 Funnel Sizes for Dry-Column Chromatography

Funnel diameter (mm)	Funnel length (mm)	Weight of silica (g)	Weight of sample (mg)	Fraction size (cm³)
30	45	15	15–500	10–15
40	50	30	500–2000	15–30
70	55	100	1000–5000	20–50

2. Under vacuum, preelute the column with a solvent that will give a TLC R_f of about 0.2 for the least polar constituent of the sample. If the silica has been packed correctly, the solvent should run down the column with a horizontal front; but if it channels, the column should be sucked dry and repacked. Keep the surface of the silica covered with solvent while preeluting until the solvent starts collecting; then suck it dry.

3. Load a solution of the sample using the same solvent as the one used for preelution in an even layer onto the surface of the silica. If the sample is insoluble in the preelution solvent, it can be preadsorbed onto a small quantity of silica. This is then spread on the surface of the silica in the funnel.

4. Elute the column by sequentially adding solvent fractions to the funnel according to the quantities indicated in Table 11.1, sucking the silica dry in between each fraction and keeping the fractions separate. For each successive fraction, increase the solvent polarity by increasing the proportion of the more polar solvent by about 5%–10% (e.g., from 50% EtOAc/50% petrol to 55% EtOAc for a 10% increase). Analyze the fractions by TLC to determine the locations of the components of interest. As a rough guide, the solvent mixture that would give the compound a TLC R_f value of about 0.5 will probably elute it.

As with any chromatographic technique, expertise will come only with experience; but given that, you should be able to separate quite closely running compounds quickly using this technique.

11.7 Preparative TLC

There was a time when preparative TLC was used routinely for separating small quantities of compounds of similar polarity. However, this technique has largely been superseded by the use of flash chromatography and HPLC, both of which can give better resolution than the former. Thickly coated "prep TLC plates" should definitely be considered a thing of the past, but large (20 × 10 cm) analytical TLC plates can prove useful as prep TLC plates in some instances. These will typically separate about 10 mg of compounds.

To perform preparative TLC, you first need to evenly spot the mixture along the bottom of a TLC plate. A high degree of skill is required to draw a TLC dropper across a line at the base of the plate applying a very thin and even line of the sample solution without damaging the silica. In between each application, the solvent is allowed to dry before a further application is made; this process is repeated until all the sample has been loaded.

After running the plate (multiple elution is often required for good separation), it needs to be viewed in a nondestructive way. This is best done using a UV lamp; consequently, it is very difficult to use prep TLC for non-UV-active compounds. The position of the bands should be marked with a fine pencil. A sharp scalpel can then be used to cut the bands and scrape them carefully off the plate. The compounds can then be separated from the silica by washing with a very small quantity of methanol and then filtering off the silica through a cotton wool plug in a Pasteur pipette. After evaporation, it is usually necessary to redissolve the compound in dichloromethane and filter it again to remove the traces of silica that have dissolved in the methanol. Great skill is required to run preparative TLCs effectively.

11.8   Medium pressure and prepacked chromatography systems

A number of chromatography systems are now available that fall between simple flash systems, as described in Section 11.5, and preparative HPLC, which is described in Section 11.9. These systems can be loosely referred to as medium pressure liquid chromatography (MPLC) systems. A wide range of commercial automated chromatography systems of this type are available, for example, the Biotage Isolera and Fashmaster systems and the Teledyne Isco Combiflash system. Alternatively, simple systems can be assembled from readily available components (Figure 11.22). The simplest version of this setup has a solvent pump connected to a chromatography column, with a three-way tap in line between the pump and the column to allow sample injection. The fractions are then collected in test tubes and analyzed by TLC in the same way as in flash chromatography. A setup of this type requires a solvent pump that operates up to 100 psi with a controllable flow rate of 6–100 cm^3/min. It is important to ensure that the tubing and the column can withstand pressures of 100 psi. More sophisticated versions of this setup can include solvent pumps that provide gradient solvent mixing, a sample injector port, an in-line detector, and a fraction collector.

The components of simple MPLC systems are normally connected using PTFE tubing, which is extremely versatile. The tubing is easily cut to size and can be connected to the various components of the system using Luer fittings and plastic screw ferrules (Figure 11.23).

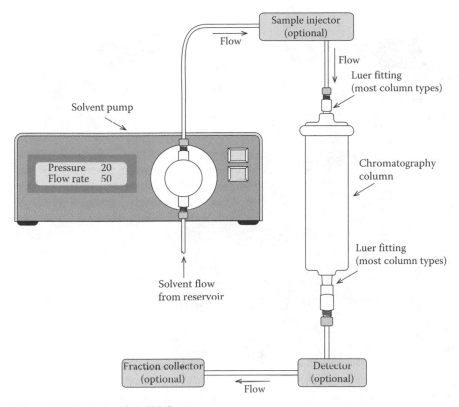

Figure 11.22 A simple MPLC system.

A wide range of chromatography columns that are compatible with MPLC can be purchased. They usually have screw tops and Luer fittings to facilitate connection to the MPLC system. They can be purchased empty or prepacked and are usually constructed of solvent-resistant plastic. A 50 × 5 cm column is ideal for separations on an 8–20 g scale. For most purposes, ordinary flash silica (40–60 μm) can be used, but finer silica (e.g., 15–20 μm) can provide higher resolutions for difficult separations. It is important to note that back pressure increases as silica particle size decreases, so lower flow rates may be necessary if finer silica is used.

When using prepacked columns, the simplest way to introduce the sample is to inject a solution onto the top of the column from a syringe connected to the Luer connector. The pumping system can then be connected to the top of the column and the solvent flow turned on. In order to introduce a sample to the column without disconnecting the column from the solvent flow, a sample injector can be incorporated into the system. A typical injector valve configuration (Rhodyne type) is shown in

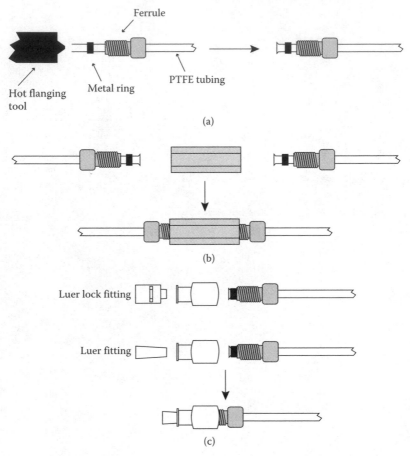

Figure 11.23 Using ferrule and Luer connections: (a) fitting a ferrule to 3 mm PTFE tubing, (b) connecting 3 mm PTFE tubing using a threaded sleeve, and (c) Luer connectors enabling PTFE tubing to be connected to Luer fittings.

Figure 11.24. The sample is injected in the "load" mode, with the injector loop is isolated from the solvent flow. With solvent flowing from the pump, the injector valve is then switched to the "inject" mode and the solvent stream is diverted through the load loop, introducing the sample onto the column. Sample load loops can be made from a length of PTFE tubing with plastic screw ferrules at either end. The sample loop can be kept coiled for tidiness, and it is useful to have a number of them of different lengths (volumes) at hand. Commercial automated chromatography systems usually incorporate some kind of sample injector.

Because the solvent flows rapidly (6–100 cm^3/min) through an MPLC system, care needs to be taken when collecting fractions from the column. Test tubes will fill rapidly, and any delay in swapping test tubes can lead

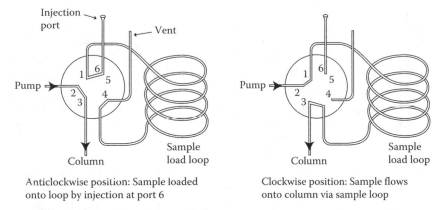

Figure 11.24 Flow through an MPLC injection valve.

to significant spillage of the eluent from the column. An automated fraction collector can be very useful here, but it is essential to use test tube racks that hold the tubes very close together to minimize spillage as the collector changes fractions. It is important to remember that when working on a larger scale, larger fractions can be collected without losing resolution. If your column gives poor resolution of the components, collecting a lot of small fractions will not improve separation; it will simply leave you with a lot of fractions containing mixtures. So save yourself time and collect large fractions from large columns.

In most cases, the fractions collected from MPLC are analyzed by TLC or analytical HPLC, but it is also possible to incorporate a detector (e.g., UV or refractive index) into the system in the same way as in HPLC. Commercial automated MPLC systems usually incorporate a detector and can be programmed to switch fractions according to a response from the detector. Although this can be convenient, it does not improve the effectiveness of the separation.

11.9   Preparative HPLC

11.9.1   Equipment required

A preparative HPLC system has exactly the same components as an analytical HPLC system (Chapter 9, Section 9.3.2); the only difference is that some components are larger in the former. With some systems, you can simply switch between analytical and preparative modes. With others, you may need to change pump heads, solvent load loops, and analytical flow cells to allow the higher volumes and flow rates required for preparative HPLC. For preparative HPLC, the injection valve will need to be the large bore type (e.g., Rheodyne 7125) fitted with a large load loop (up to 5 cm³) and, of course, the columns are much larger. It is also useful, but not essential, to have a fraction collector linked to the detector so that fractions are only collected when the detector shows that a peak of interest is being eluted. Many sophisticated, computer-controlled systems are available now, including ones that can be set to inject samples automatically and collect selected peaks into designated flasks. Thus, large quantities of material can be separated over multiple injections in an automated fashion. However, for most research purposes the prime requirement is for one-off separations.

The choice of detector can be quite critical. UV detectors are very sensitive but are of limited use for molecules without a UV chromophore. An evaporative light-scattering detector (ELSD) will detect non-UV compounds, but it is not universally applicable, so you will need to determine the best means of detection before starting.

11.9.2 Running a preparative HPLC separation

The sequence of events for running a preparative HPLC separation is essentially the same as that for running an analytical HPLC separation (Chapter 9, Section 9.3.2). However, there are a few significant differences between the two that should be taken into account:

1. Care of preparative HPLC columns: *Preparative HPLC should be used only for the separation of clean mixtures.* Some other form of chromatography should always be carried out first to remove any material that is significantly more polar than the compounds to be separated, and any suspended solid also should be removed. Preparative HPLC is a very powerful technique, and it is often possible to separate compounds that do not separate on TLC. However, preparative HPLC columns are extremely expensive and, although they can last for a very long time, they can easily be ruined by one thoughtless application. Tiny solid particles will damage a column and, for this reason, sample solutions should *always* be passed through a fine filter directly before loading.
2. Choice of column size, solvent, and flow rate: It is very difficult to draw a correlation between the behavior of a mixture on TLC and the way it runs on HPLC. It is therefore essential to carry out a few analytical HPLC runs to find an appropriate solvent system and determine the quantity that can be separated. You should use the same type of HPLC column for both analytical and preparative work. Most manufacturers provide sets of columns that run from microanalytical sizes to large preparative scale sizes and have uniform characteristics through the series. A frequently encountered problem with preparative HPLC columns is the packing down of

the solid phase, which occurs with constant use and causes voids, particularly at the ends of the column. For this reason, HPLC columns that are fitted with a compression joint that allows the voids to be closed are preferred. It is also advantageous to use guard columns in conjunction with all HPLC columns to prolong their lifetime.

The sizes of the columns increase in regular steps, and it is therefore very easy to correlate between the smaller analytical columns and the larger preparative columns. The size of columns commonly used by organic chemists for analytical work is 4.5 mm × 25 cm. This is a useful size for method development as it takes only a few minutes to equilibrate and run such columns. A column of this size can also be used as a small-scale preparative column for quantities up to 40 mg. For normal analytical HPLC, the intention is to develop a method that gives the best peak shape with good separation and, therefore, column overloading is to be avoided. However, a good preparative HPLC method is one that separates the components and allows you to '"get away" with maximum overloading, peak shape being largely irrelevant. For full-scale preparative work, either a 22.5 mm or a 45 mm width column is normally used. It is possible to scale up to this size knowing the conditions used for the smaller column. If the columns are of the same length, the scale-up factor will depend on the area of the column surface; thus,

$$\text{Scale-up factor } (F) = \frac{\text{Area (large column)}}{\text{Area (small column)}} = \frac{\pi r^2 (\text{large column})}{\pi r^2 (\text{small column})}$$

$$= \frac{r^2 (\text{large column})}{r^2 (\text{small column})}$$

Using this equation, the scale-up factor between a 4.5 mm width column and a 22.5 mm width column is 22 and that between a 22.5 mm width column and a 45 mm width column is 4. If the columns are not of the same length, then simply multiply the factor by the proportionate length difference. For example, going from a 4.5 mm × 12.25 cm column to a 22.5 × 25 cm column, the scale-up factor would be 45. To work out the conditions for running the larger column, simply scale the flow rate and the quantity of sample used on the small column by the scale-up factor; this should produce identical results on the large column. As a rough guide, a 4.5 mm × 25 cm column will run at about 0.75 cm^3/min and can be loaded with 5–35 mg, a 22.5 mm × 25 cm column will run at about 16 cm^3/min and can be loaded with 100–800 mg, and a 45 mm × 25 cm column will run at about 64 cm^3/min and has a capacity of about 450 mg to 3.2 g.

All the solvents used for preparative HPLC must be of very high grade and must be filtered and degassed before use. Many HPLC machines have a degasser incorporated into the instrument. If not, degassing for HPLC can be achieved by bubbling helium through the solvent via a gas diffuser, by standing the bottle of solvent in an ultrasonic bath, or by stirring the solvent under vacuum. The solvent inlet line of the HPLC system should always incorporate a filter to prevent small particles from entering the system. Diethyl ether is generally not a useful solvent for HPLC work as it is best to use a system where the more polar constituent is also the least volatile constituent.

References

1. W.C. Still, M. Khan, and A. Mitra, *J. Org. Chem.*, 1978, **43**, 2923.
2. L.M. Harwood, *Aldrichimica Acta*, 1985, **18**, 25.

chapter twelve

Small-scale reactions

12.1 Introduction

For the purposes of this chapter, we will define small-scale reactions as those involving reaction mixture volumes of less than 5 cm^3. When performing organic reactions on this scale, special problems arise, most notably the following:

1. Difficulties in measuring out small quantities of sensitive reagents.
2. Significant losses of material due to apparatus design.
3. Difficulties in excluding trace amounts of water from moisture-sensitive reactions.

Whenever reactions are performed, material losses are obtained as a consequence of the preceding problems. Normally, these losses only account for a few percent of the total material; however, this percentage can increase dramatically as the reaction scale decreases. For example, if a moisture-sensitive reaction was carried out on a 1-mole scale, it would take 18 g of water to completely stop the reaction occurring, but if the same reaction is carried out on a 0.1-mmol scale, then only 1.8 mg of water would completely quench the reaction. A highly skilled organic chemist should be able to successfully carry out moisture-sensitive reactions on a 0.01-mmol scale.

The problem of weighing out small quantities of sensitive reagents is best solved by accurately weighing larger quantities and making up solutions in an inert solvent, ideally the reaction solvent. Since the molarity of the solution is known, a quantitative aliquot of this solution can then be added to the reaction mixture using a syringe. This effectively reduces the problems of weighing out the material to those that exist in larger-scale reactions, discussed in Chapters 9 and 13. As a general rule, many of the techniques used in setting up reactions discussed in the Chapter 9 can, with care, be applied to small-scale reactions. Indeed, there are a variety of miniature chemical apparatus commercially available for just this purpose.

This chapter will outline some of the more specialized techniques that can be employed to alleviate the problems associated with small scale and that can be carried out without the requirement of relatively expensive specialized glassware.

12.2   Reactions at or below room temperature

When carrying out small-scale reactions at or below room temperature, it is quite possible to use a conventional apparatus setup. Most reactions are carried out in round-bottom flasks, which are available in sizes down to 1 cm³. However, problems arise when you come to work up the reaction mixture. If an aqueous or organic extraction is required, then the material must be transferred to a separating funnel. This will inevitably lead to some loss of sample during the transfer, but this should be minimal. The main problem arises from the fact that separating funnels rarely come in sizes below 10 cm³. Also, because of their design, there will always be some loss of material in the apparatus itself. This is mainly because the reaction mixture necessarily comes into contact with a large surface area of glassware, and retrieving material coated over this large area tends to be difficult. A useful alternative for extractions involving 2 cm³ or less is to use a glass sample vial or small test tube (Figure 12.1). The reaction mixture can be washed into the sample vial, and the extraction solvent added. The lid of the vial is then put on

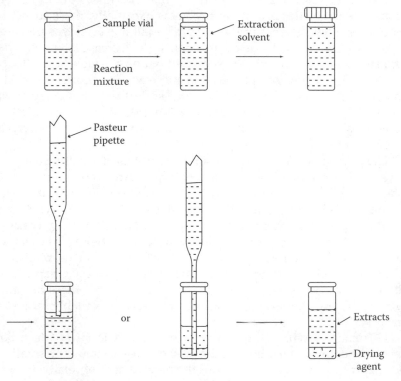

Figure 12.1 Small-scale work-up in a sample vial.

and the mixture shaken. Removal of the lid allows any pressure buildup to be released. Particular care should be taken when removing the lid to avoid spillage.

The required solvent layer is then removed using a Pasteur pipette. Three or four extractions are usually sufficient to recover the majority of the material. This technique is particularly straightforward for diethyl ether extractions of aqueous solutions where the ether layer is required, since it can readily be pipetted from the top of the aqueous layer (Figure 12.1). If you require the lower solvent layer, this can be recovered either by pipetting away the top layer or by pipetting from the bottom of the vial. Because efficient separation of the two phases is required, it is preferable to use a tall, thin vial rather than a short, fat one. A second vial containing drying agent can then be used to dry the extracts.

Removal of the drying agent is normally achieved by filtration of the solvent mixture. On a small scale, this is best achieved using a Pasteur pipette fitted with a cotton wool plug as the filtration apparatus (Figure 12.2). Once the solvent has been transferred into the filtration pipette, it can be forced through the plug by applying pressure with a pipette teat. Evaporation of the solvent in the normal way then yields the crude reaction product.

As stated earlier, material is inevitably lost with each transfer of apparatus. It is possible to cut down the number of transfers by using the sample vial as the reaction vessel. Very small magnetic fleas are commercially available and will fit most small sample vials. Consequently, with magnetic stirring, the vials can serve as small reaction vessels (Figure 12.3). They are conveniently attached to the top of a magnetic stirrer machine using plasticine.

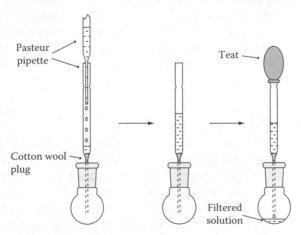

Figure 12.2 Small-scale filtration using a plugged Pasteur pipette.

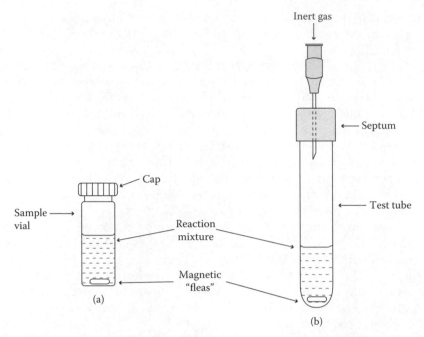

Figure 12.3 Use of sample vial or small test tube as a reaction vessel.

When capped, the vial is a sealed system (Figure 12.3a); conse-quently, this setup is only useful for small-scale reactions at room tem-perature that do not involve an increase or decrease in pressure inside the reaction vessel. For reactions at low temperature, or those requiring a positive pressure of an inert gas atmosphere, it is often more conve-nient to use a Pyrex test tube fitted with a septum (Figure 12.3b). In all other respects, the arrangement is the same, and since Pyrex test tubes are available in a range of sizes, this apparatus can cope with a range of reaction volumes down to about 0.2 cm^3. Again, the reaction vessel can also be used for a subsequent extraction procedure simply by replacing the septum with a cap.

12.3   Reactions above room temperature

Carrying out small-scale reactions above room temperature, particularly those involving solvent at reflux, can be difficult. The main problems are associated with preventing the evaporation of very small amounts of solvent and material losses in the apparatus. This is usually caused by losses of material through ground-glass joint attachments of con-densers to the reaction flask and by inefficient condensation of the sol-vent vapor. One solution is to use a sealed tube (see Section 9.4.2) as the

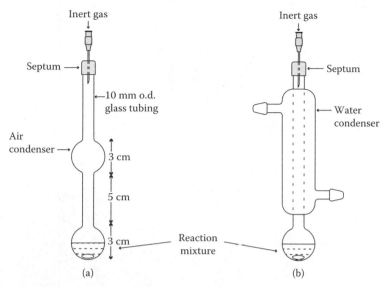

Figure 12.4 Small-scale air condenser and water condenser systems: (a) air condenser and (b) water condenser.

reaction vessel, and this is the equipment of choice for reactions involving volatile solvents (≤100°C) on scales below 1 cm³. For higher boiling solvents, it is usually possible to use a one-piece apparatus, incorporating an air condenser system (Figure 12.4a). This apparatus can be conveniently constructed at the required size from a piece of Pyrex tubing. The air condenser is adequate enough to prevent the evaporation of higher boiling solvents (≥100°C). Alternatively, a one-piece apparatus incorporating a water condenser can also be used (Figure 12.4b). In this case, the condenser system is more efficient, but construction of the apparatus is correspondingly more complex. In both cases, the reflux apparatus can be constructed to allow reaction volumes down to about 0.5 cm³.

12.4   Reactions in NMR tubes

In many cases, it is necessary to monitor closely the progress of a small-scale reaction by methods other than TLC or HPLC. One very useful alternative is NMR. With larger-scale reactions, this can be done by removing an aliquot of the reaction mixture and recording its NMR spectrum. Obviously, this approach cannot be applied to reactions involving relatively small quantities of material. The answer is to carry out the reaction in a NMR tube. NMR tubes can conveniently be used as reaction vessels (Figure 12.5) in the same way as test tubes, but without magnetic stirrer bars.

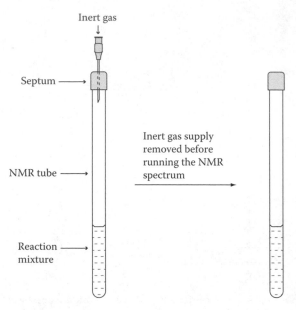

Figure 12.5 Reactions in NMR tubes.

The reaction can then be monitored by recording the spectrum of the reaction mixture directly. There are however several important points to note when using NMR tubes as reaction vessels:

1. The reaction solvent should be chosen carefully to ensure that it does not obscure the NMR region to be observed.
2. Magnetic stirring cannot be used because the magnetic flea in the NMR tube would interact with the NMR field causing severe broadening of the spectrum. Similarly, the presence of paramagnetic material in the reaction mixture will lead to broadening of the spectrum.
3. For high-quality NMR spectra to be recorded, the reaction mixture should be homogeneous. The presence of solids in the reaction mixture will lead to poorly resolved spectra.

As a general rule, you should use the reaction solvent that is normally employed for the type of reaction being carried out. Make sure to use fully deuterated solvent if ^{1}H NMR spectra are to be observed. Agitation of the reaction mixture is probably best achieved using sonication in an ultrasonic bath (Section 9.6.4), although periodic shaking will often suffice.

If you need to carry out the reaction at elevated temperatures, then it is usually advisable to use a sealed NMR tube. Thick-walled NMR tubes are commercially available and will withstand increased reaction pressures typical of those obtained in sealed tube experiments.

12.5 *Purification of materials*

The purification of small quantities of materials (≤50 mg) also poses certain problems. A number of simple techniques used are outlined in Sections 12.5.1 to 12.5.3.

12.5.1 *Distillation*

By far, the most convenient method of carrying out distillation on a small scale is to use a Kugelrohr apparatus (Figure 11.14). To cut down on losses through ground-glass joint connections, it may be beneficial to use a one-piece Kugelrohr bulb set (Figure 12.6). This can be conveniently made to the size required from a piece of Pyrex tubing.

After distillation is complete, the apparatus is left to cool, and the purified material is recovered by cutting up the apparatus into three sections using a glass knife.

12.5.2 *Crystallization*

Crystallizations on a small scale are most conveniently carried out using a Craig tube apparatus as described in Section 11.2.2.

12.5.3 *Chromatography*

All the normal chromatography techniques (see Chapter 11) can be used to purify small quantities of material. Indeed, preparative HPLC is often more successful when small quantities of material are involved.

In the case of flash chromatography, it is often impractical to simply scale down the equipment. A useful alternative is to employ a Pasteur

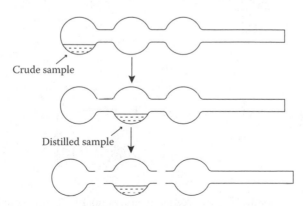

Crude sample

Distilled sample

Figure 12.6 One-piece Kugelrohr bulb set.

pipette as the column. Such a column is easily constructed using a pipette fitted with a cotton wool plug (Figure 12.7a). The pipette is then filled with the required adsorbent (typically silica gel). Do not fill the pipette more than three-quarters full, otherwise there will be insufficient room for the solvent. To prepare the column, dry silica is added to the pipette. The eluting solvent is then added to the top of the column and allowed to run through the column under gravity. More solvent is added to the top of the column as required. Once the solvent starts to appear at the bottom of the pipette, pressure can be applied using a pipette teat, forcing the solvent through at a faster rate. After about two column volumes of solvent have been passed through, all the air should have been removed from the silica and the column should be ready for use. The sample is applied to the top of the column in the usual way and pressure is applied using a pipette teat. The main difference between this arrangement and the more usual flash chromatography setup is that the pressure applied to the top of the column is not constant. The teat is continuously being removed to allow the addition of more solvent to the top of the column. Consequently, the solvent

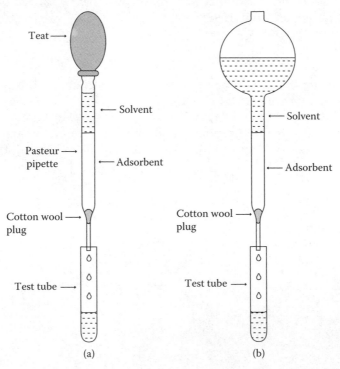

(a)   (b)

Figure 12.7 Small-scale flash chromatography: (a) miniature column using a Pasteur pipette and (b) miniature column with a built-in solvent reservoir.

is not passing through the column at a constant rate. In most instances, this does not appear to significantly affect the separations achieved. However, if this proves to be a problem, a miniature chromatography column with solvent reservoir can easily be constructed from Pyrex glass tubing (Figure 12.7b). With a cotton wool plug in the bottom, this can be used in exactly the same way as the larger columns described in Section 11.5.2.

chapter thirteen

Large-scale reactions

13.1 Introduction

This chapter deals with some of the techniques required when working with larger scale reactions. For the purposes of this chapter, we will define a larger scale reaction as one involving reaction volumes between 1 and 5 L. Working on reaction volumes in excess of this usually requires the use of specialist equipment and is beyond the scope of this book.

Compared with small scale, there are additional issues that must be considered when moving to large scale work. Safety is the main concern because on a larger scale, all the hazards associated with a reaction are increased. It is, therefore, essential to test the reaction on a small scale before attempting it on a larger scale and always limit any scale up to ≤10-fold.

Some of the main issues that need to be considered when working on a larger scale are as follows:

1. Everything will tend to take longer. Because of the larger volumes involved, reagent addition, evaporation of solvent, phase separation during work-up, and so on, will all take longer than they did on a smaller scale, so make sure that you have enough time available to complete the experiment. The reaction apparatus will also be significantly heavier, so make sure that it is held securely.
2. Effective agitation can be a problem. Mechanical stirring is usually necessary as magnetic stirring is ineffective in larger equipment. It can also be beneficial to include a baffle in the reaction vessel as this will improve the effectiveness of the agitation. Something as simple as a thermometer dipping into the reaction mixture will cause significant turbulence and have a beneficial effect on mixing.
3. Addition of reagents will require careful planning. Manual addition via a syringe can be used for smaller quantities (<50 cm^3), but syringe pumps are more useful for addition of larger quantities, particularly when slow, controlled addition is required. Addition of solutions via pressure-equalized dropping funnels can also be used, but these do

not provide particularly good control of addition rate. A variety of pumping devices are also available for controlled addition of larger volumes.

4. Control of the reaction temperature becomes more difficult. It takes much longer to heat up or cool down larger reaction volumes. The internal temperature of larger round bottom flasks is poorly controlled by simple heating and cooling baths. Jacketed vessels are recommended for reactions above or below ambient temperature.

5. Exothermic reactions can prove to be a particular challenge on a larger scale. On a small scale, any heat generated during a reaction is usually dissipated quickly because the reaction mixture has a relatively large surface area compared to its volume. However, the relative surface area of larger reactions is much smaller and so heat cannot dissipate so easily. Any buildup of heat will cause a faster rate of reaction and this will generate even more heat, and so on. If this happens, the reaction will quickly run out of control. Thus, when planning a large-scale reaction, it is essential that you ensure this cannot happen. Before scaling up, always monitor the reaction on a small scale and measure any exothermic behavior. A common mistake with an exothermic reaction is to add all the reagents together at the start or to add one reactant to another quickly, resulting in a buildup of reactive components. There is often an initial delay (induction period) before heat starts to be produced. If you are not careful, too much reagent may have been added by this point and it will then act as fuel to produce more heat until it has been consumed. To carry out an exothermic reaction under controlled conditions, it is best practice to bring the reaction mixture to the desired temperature before adding the limiting reagent. Then add the reagent at such a rate as to maintain the reaction safely at that temperature. Starting an exothermic reaction at an elevated temperature might seem counterintuitive, but if the reaction needs to be performed above room temperature, this is often the safest way to avoid build up of the reagent as it is being added. Again, a jacketed vessel connected to a heater–chiller circulator provides a convenient way to maintain a reaction mixture at constant temperature during this kind of addition.

6. Some aspects of isolation and purification can be more challenging. For example, considerably more time should be allowed for phase separations and solvent evaporation. It is also less convenient to purify large quantities of material by chromatography.

Some issues also become less significant on a larger scale, for example, moisture-sensitive reactions can be easier to perform because traces of water entering the system will only account for a small percentage of the reaction mixture. A convenient way to keep a large reactor dry is to add the solvent before other components and distill a small amount of solvent off to remove water as an azeotrope. This works best for solvents that form good azeotropes, such as toluene and chloroform, but can be effective with other solvents as well. Material losses that occur during isolation of the product also become less significant when working on a larger scale. It is also often easier to crystallize or distill materials on a larger scale.

13.2   Carrying out the reaction

Although larger versions of standard laboratory equipment are often adequate for larger scale reactions (e.g., Figures 13.1 and 13.2), jacketed vessels (Figures 13.3 and 13.4) can offer significant advantages.

13.2.1   Using standard laboratory equipment

A standard setup for a larger scale reaction that needs to be heated is shown in Figure 13.1. The pressure-equalized dropping funnel can be filled by removing the stopper and pouring in the required material or, if the material is moisture sensitive, then it can be transferred into the dropping funnel via a cannula by replacing the stopper with a septum. Heat can be supplied via a heating bath. It is recommended that any heating source be mounted on a laboratory jack so that it can be rapidly removed in case of emergency.

A similar setup (Figure 13.2) can be used for larger scale reactions that need to be performed at or below room temperature. If the solution to be added from the dropping funnel requires cooling prior to addition, it is possible to use a jacketed dropping funnel, with the cooling mixture (e.g., dry ice/acetone) placed in the jacket.

13.2.2   Using a jacketed vessel

A jacketed vessel connected to a heater–chiller circulator will generally allow better control over the reaction conditions on a larger scale. It can also often be useful to have two coupled jacketed vessels. For example, a second jacketed vessel can be used to pre-equilibrate a reagent at a specific temperature before it is transferred to the main reaction vessel. Typical setups for jacketed vessels are shown in Figures 13.3 and 13.4.

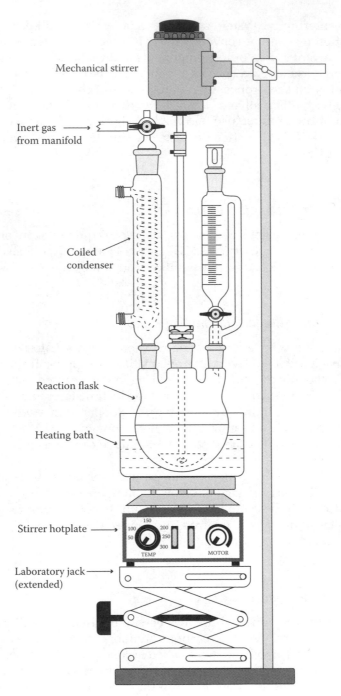

Mechanical stirrer

Inert gas
from manifold

Coiled
condenser

Reaction flask

Heating bath

Stirrer hotplate

Laboratory jack
(extended)

Figure 13.1 Large-scale reaction setup (heat).

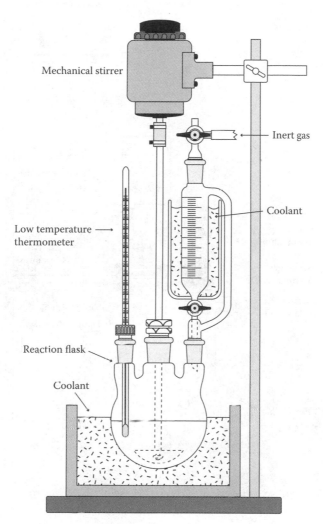

Figure 13.2 Large-scale reaction setup (cool).

A range of automated reactor systems are also available for use on larger scale reactions, and a brief description of these can be found in Sections 4.5 and 4.6.

13.3   Work-up and product isolation

The principles of work-up and product isolation on larger scale are the same as for other scales (Chapter 10). However, because chromatography on a large scale is difficult, you should spend some time designing an

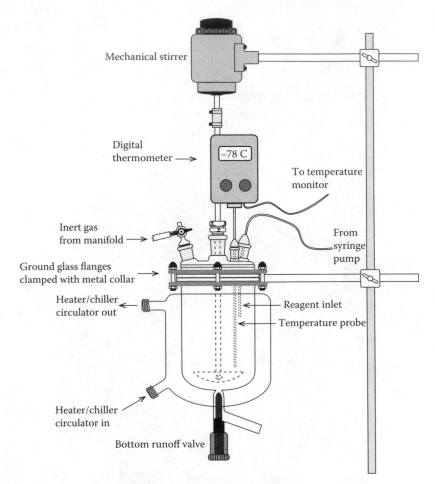

Mechanical stirrer

Digital
thermometer →   −78 C

To temperature
monitor

Inert gas
from manifold →

From
syringe
pump

Ground glass flanges
clamped with metal collar →

Heater/chiller
circulator out

Reagent inlet

Temperature probe

Heater/chiller
circulator in

Bottom runoff valve

Figure 13.3 Large-scale reaction—jacketed vessel with syringe pump (cold).

optimal work-up procedure before starting the reaction. On a smaller scale, it is often sufficient for work-up to simply remove inorganic residues and solvents prior to chromatography, whereas on a larger scale, it is essential to use the work-up to remove as many impurities and by-products as possible prior to purification.

It is also essential to take into account the reactivity of the reaction mixture and the physical scale of the procedure. Reaction mixtures that are likely to react vigorously with water (e.g., those involving Grignard reagents, LiAlH$_4$, etc.) are best quenched by adding the reaction mixture, in a controlled manner, to a quench solution at low temperature. For aqueous work-ups, it is often helpful to include crushed ice in the quench vessel. When using jacketed vessels, a second jacketed vessel held at an

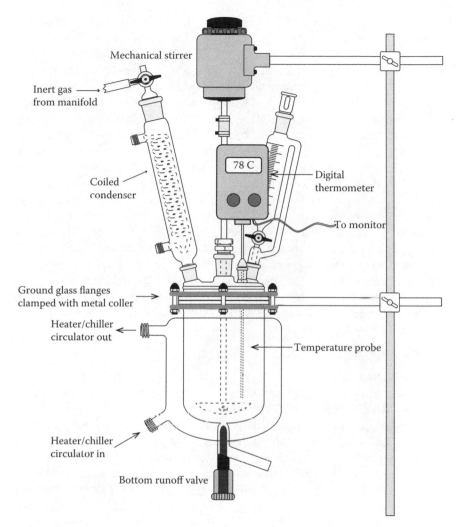

Figure 13.4 Large-scale reaction—jacketed vessel (heated).

appropriate temperature can be used for the quench. An advantage of using jacketed vessels (with bottom runoff valves) is that they are more effective for large-scale separations than using a separating funnel. Large separating funnels can be difficult to handle, and care must be taken to ensure that pressure does not build up inside, otherwise the contents may be expelled violently. Holding the separating funnel in a manner such that pressure can be regularly released from the tap is crucial. With two linked jacketed vessels these issues are avoided because the setup can be open to the atmosphere and because the vessels are permanently clamped their physical weight is not a problem.

Before designing a work-up procedure for a large-scale reaction, it is essential to write out a balanced equation for the reaction. This should include all reactants, products, reagents, catalysts, by-products, and any other impurities. Any knowledge you have about the relative amounts of these components will be very useful, but understanding their individual physical properties is the key to designing an effective work-up. Any tests that you are able to do to estimate the relative solubility of the components of the reaction mixture will be useful in designing the work-up. The difference in physical properties between acids, bases, and neutral compounds is particularly easy to exploit, but less extreme polarity differences can also be used to separate components. Once you have a good understanding of the components of a reaction mixture, it should be possible to devise the best work-up procedure. Partitioning components between aqueous and organic phases is particularly convenient and can be used on any scale. This technique is especially useful for separating salts from neutral components or a polar, aqueous soluble compound from nonpolar ones. Directly precipitating a solid product from the reaction mixture is another option and is often used on a large scale. More details on how to design and implement these work-ups can be found in Chapters 10 and 11.

13.4   Purification of the products

Although chromatography, including separation of enantiomers, can now be carried out on an industrial scale, it is always expensive and should be avoided where possible. The best methods for final purification on a larger scale are fractional distillation (for liquids or oils) and recrystallization (for solids). A detailed discussion of these techniques can be found in Chapter 11. If you plan to use distillation to purify a product on a larger scale, it is essential that you first check the thermal properties of the compound and of any potential distillation residues. Many commonly encountered organic materials are highly explosive (e.g., peroxides, ozonides, azides, diazo compounds, skipped alkynes, etc.) and should never be heated. Even if they are not explosive, most organic compounds will decompose on heating above 150°C, especially if this is done in the presence of oxygen.

chapter fourteen

Special procedures

14.1   Introduction

This chapter deals with some of the specialized procedures that might be encountered. All the topics covered here have one thing in common, namely that the particular type of apparatus used will vary from one laboratory to another and accordingly detailed instructions will not usually be given here for any one piece of apparatus. Representative systems are shown, and common operating procedures discussed. Often there will be someone in the department who is responsible for, or has particular expertise in, one of these techniques. If this is the case, always consult this person before attempting the reaction.

14.2   Catalytic hydrogenation

Caution: Extreme care must be taken whenever hydrogen gas and active catalysts are used. Observe local safety precautions strictly.

If you are unfamiliar with catalytic hydrogenation or with the particular local apparatus, do find someone with experience with the apparatus and technique and familiarize yourself with the manipulations and precautions before attempting the reaction.

The reduction of organic compounds using hydrogen and a metal catalyst is a reaction that is often encountered. Most catalytic hydrogenations are carried out at atmospheric pressure, and organic chemistry laboratories will have their own "atmospheric" hydrogenation apparatus. These consist of a gas burette (or burettes) connected to a hydrogen supply and to the reaction flask. The detailed operation will depend upon the precise equipment used, and a written step-by-step procedure should be available for the setup. The underlying principle behind most atmospheric pressure hydrogenators is the same, and a typical apparatus arrangement is shown in Figure 14.1. Other apparatus might be arranged somewhat differently, but the operating procedure for most atmospheric pressure hydrogenators is the same. For the apparatus shown, the procedure is as follows:

1. With taps A and B open, raise the leveling bulb to fill completely the gas burette with liquid (usually water or aqueous copper(II) sulfate, but sometimes mercury), then close tap B.

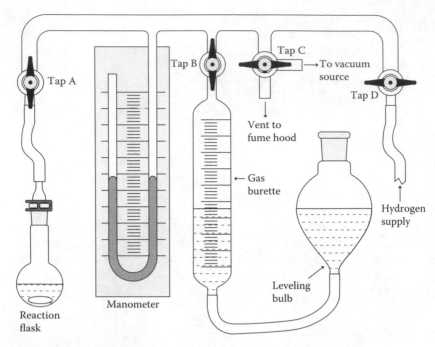

Figure 14.1 Schematic diagram of an atmospheric hydrogenator.

2. Connect your reaction flask using vacuum tubing and close tap A. See below for important information on preparing the reaction mixture. It is a good idea to use a long-necked reaction flask in case the solvent bumps, and a thin ring of grease should be applied to seal around the top of the connecting joint.

3. Connect the hydrogen supply but keep tap D closed. Then connect the vacuum source to tap C and open tap C carefully to evacuate the main part of the system. Once the system has stabilized (indicated by the manometer), close tap C and carefully open tap A partially to evacuate the reaction flask. Now with the hydrogen supply turned on, carefully open tap D to introduce hydrogen.

4. To saturate the system with hydrogen, repeat step 3 twice more.

5. The system should now have taps A and D open and tap C closed, with the hydrogen supply on. Now slowly open tap B to allow the gas burette to fill with gas, adjusting the leveling bulb as necessary. More than the theoretical amount of hydrogen should be introduced if possible.

6. With the gas burette filled, close tap D, adjust the leveling bulb so that the manometer mercury levels are equal, and record the burette reading.

7. The reaction is then agitated (stirred or shaken) and the uptake of hydrogen monitored using the burette. To measure the amount of gas that has been absorbed, equalize the manometer mercury levels before recording the burette reading.
8. When the required amount of hydrogen has been used, or no more is being taken up, purge the system with nitrogen (follow step 3 using nitrogen instead of hydrogen) and then remove the reaction flask.

It is important to follow the procedure closely and to consider the effect of opening any tap before doing so.

When filling the reaction flask, add the catalyst first, followed by the solvent, and then your substrate. Addition of an active batch of catalyst to solvent can cause fires. Do make sure that all air is removed from the system by following the detailed procedure that applies to your particular system, and when the reaction is over, ensure that as much residual hydrogen is removed as possible; again, follow the procedure. Ensure that a safety screen is used throughout the whole operation.

Filtration of the reaction mixture must be done carefully. Filter through a pad of Celite in a sintered glass funnel but do not allow the catalyst to dry out. Wash through with more solvent and dispose of the wet catalyst/Celite mixture properly. Most laboratories have a special bottle for catalyst residues; use it. Numerous fires have resulted from wet catalyst residues being placed in a waste bin since the residue dries out in the air and ignites. Raney nickel residues are notorious in this respect.

An alternative method for removing the catalyst is to centrifuge the reaction mixture and pipette off most of the supernatant. Resuspend the residue by adding more of the reaction solvent and centrifuge again. This process can be repeated two or three more times as necessary. Once you are certain that the entire product has been extracted from the catalyst, the residues can be disposed of as outlined above. *Never* attempt a catalytic hydrogenation until you know what to do with the catalyst residue.

The gas burette procedure described above is sometimes inconvenient for small-scale work (it depends upon the type of hydrogenation system available). In such instances, the use of a small balloon of hydrogen connected to the reaction via a three-way tap, which is also connected to a manifold, can serve as a convenient alternative. This technique is illustrated in Figure 14.2.

Fill the balloon with hydrogen (several times to remove air) and connect it to the reaction flask via a three-way tap. Connect the three-way tap to the manifold, open to vacuum (carefully), and then to the hydrogen balloon (Figure 14.2a). Several cycles will be required to ensure that all the air in the reaction flask has been replaced by hydrogen. Then turn the tap so as to maintain a slight positive pressure of hydrogen in the flask (Figure 14.2b). The reaction can then be monitored in the usual way. When the reaction is

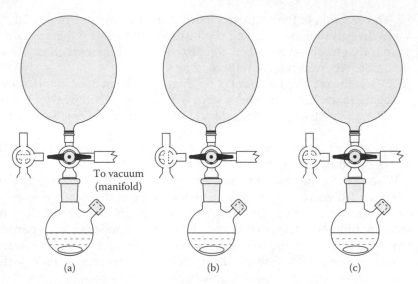

(a)   (b)   (c)

Figure 14.2 Small-scale hydrogenation using a balloon.

over, vent the excess hydrogen from the balloon (*safely*) to a fume cupboard (Figure 14.2c). Residual hydrogen can be removed from the system by the use of several cycles of evacuation followed by admission of inert gas using the manifold as above (Figure 14.2a). *All precautions referred to in the use of the atmospheric hydrogenator must also be observed here of course.*

Medium- and high-pressure hydrogenations require specialized equipment and great care. This equipment usually consists of a metal reaction vessel and the appropriate "plumbing" to allow safe pressurization with hydrogen. These reactions are potentially very hazardous and must be carried out under the close supervision of the person who is responsible for the apparatus.

14.3   Photolysis

Caution: UV radiation is very damaging to the eyes and skin.

The reactor must be properly screened before turning on the UV source. Wear special protective goggles or better still a face shield that offers protection against UV radiation if the apparatus is to be adjusted (or samples taken) while the lamp is on. When doing this also protect the hands with gloves and make sure that no areas of skin will be exposed to radiation in the event of an accident. It is much safer to turn off the lamp when such manipulations are being carried out.

Preparative photochemical reactions are usually carried out in an immersion well reactor. This apparatus is constructed from three separate components: a UV lamp, an immersion well, and a reaction flask (Figure 14.3).

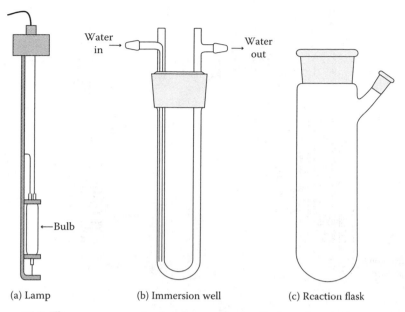

| (a) Lamp | (b) Immersion well | (c) Reaction flask |

Figure 14.3 Three components of an immersion well photochemical apparatus.

The UV lamp fits into the immersion well, and this in turn fits into the reaction flask. Once the apparatus is assembled, the reaction mixture can be added to the reaction flask via the side arm (Figure 14.4a). Air is removed from the solvent by bubbling argon through the solution or by sequentially evacuating and purging with inert gas using a double manifold attached via the side arm. It is essential to ensure that the correct choice of lamp has been made. Low-pressure lamps emit most of their radiation at 254 nm, are low power ($\leq$20 W), and require a quartz immersion well (not Pyrex). Most preparative reactions use the much higher power (100–400 W) medium-pressure lamps as these emit their radiation over a much wider range (mainly at ~365 nm with other bands at both shorter and longer wavelengths). Occasionally a filter will be required, and it is important that the correct one is used (this will be specified in the preparation being followed).

It is also possible to modify the immersion well photochemical apparatus so that it can be used as a continuous flow photochemical reactor.[1] This can be achieved by omitting the reaction flask and instead tightly coiling fluorinated ethylene propylene (FEP) tubing (1.1 mm i.d., 0.7 mm i.d.) around the immersion well (Figure 14.4b). FEP tubing is ideal for this purpose because it is solvent resistant and has excellent UV transmission properties. Up to five layers of coil can typically be used. A solution of the reactant is then pumped through the tubing and is exposed to the UV

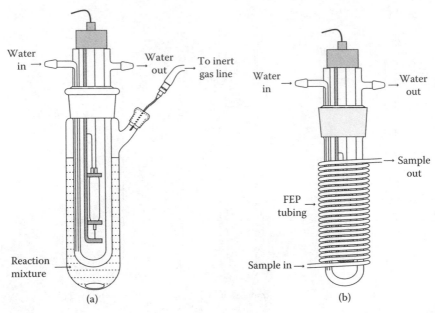

Figure 14.4 Assembled immersion well photochemical apparatus.

light while it resides in the coil. With this setup, the irradiation time can be controlled simply by adjusting the flow rate and the length of tubing used to construct the coil.

Photochemical reactions are usually run at fairly high dilution (≤0.05 M), and the solvent needs to be pure and chosen with care. It must not decompose under UV irradiation and should not absorb at the wavelength being used for the reaction. Work-up often involves no more than evaporation of the solvent followed by purification.

14.4   Ozonolysis

Caution: Ozone is toxic, and ozonides potentially explosive.

Ozone is generated using a commercial ozonator (or ozonizer), which can produce a concentration of up to 8% ozone in oxygen, and which will be available in most organic research establishments. The operation of these is very simple providing that the instructions for the particular device are followed carefully. Make sure that these are consulted before attempting the reaction.

The compound to be ozonized is dissolved in the appropriate solvent and cooled to the desired temperature. Ozone is then passed through the solution until no more starting material remains. For most purposes, an excess of ozone can be used. It can be difficult to avoid an excess, but an indicator can be added to the solution to show you when there is excess ozone in the solution and can sometimes be most valuable in avoiding overoxidation.[2]

The work-up depends upon the desired product but usually will include a reagent that reduces the ozonide. This reagent is almost always added in excess, and before any product isolation is attempted. Make sure that you allow plenty of time for the ozonide to react as isolation of ozonides is to be avoided due to their potential for violent explosive decomposition. Once the ozonide is fully reacted the reaction can be processed in the usual manner.

14.5 Flash vacuum pyrolysis (FVP)

This technique is not often encountered in synthetic organic chemistry, but it can prove invaluable in some circumstances. As with most of the topics in this chapter, the exact type of apparatus used will depend on what is available in your department. One simple (schematic) setup is shown in Figure 14.5. It is important to vaporize the substrate at the appropriate rate and to make sure that the thermolysis temperature is correct. If this information is not available, then some experimentation will inevitably have to be carried out. A typical vaporization rate might be between 0.5 and 1.0 g/h.

As with all high vacuum work, care must be taken. After all the substrate has passed through the hot tube, turn off the furnace and allow to cool to room temperature (still under vacuum). Then turn off the pump and admit nitrogen to atmospheric pressure. Remove the traps to a fume cupboard, allow to warm to room temperature, and work up in the usual way. If the desired product is unstable toward air, water, or is simply very reactive, then a more sophisticated pyrolysis system might be required, and more elaborate work-up procedures used.

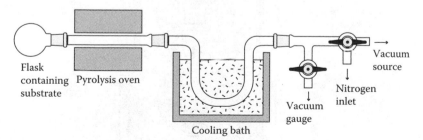

Figure 14.5 Schematic representation of an apparatus for FVP.

14.6   Liquid ammonia reactions

Caution: Ammonia is a powerful irritant, toxic, and the gas is flammable. Conduct all reactions in an efficient fume cupboard and avoid all contact with the liquid.

Liquid ammonia (b.p. −33°C) is a solvent that is not encountered frequently, but which does have several important general uses, in particular "dissolving metal" reductions ("Birch" type reductions) and most reactions involving lithium amide or sodium amide as base. Ammonia gas from a cylinder is condensed directly into the flask (Figure 14.6).

The apparatus is set up as in Figure 14.6, and a rapid flow of ammonia is used to flush out the system. A small volume of acetone (or ethanol) is poured into the condenser, and solid carbon dioxide pellets are added (very slowly at first) until the condenser is nearly full. The ammonia will begin to condense, and when the required volume has been collected in the reaction flask, the ammonia flow is shut off and the gas inlet is replaced by a septum or stopper. If undried impure ammonia will suffice, and often it will, the reaction is then carried out as normal. A cooling bath (Section 9.4.1) can be added if a long reaction time is anticipated or if a temperature below −33°C is required.

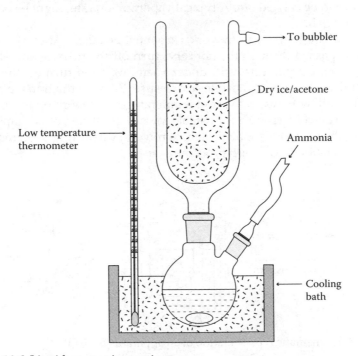

Figure 14.6 Liquid ammonia reaction.

If dry liquid ammonia is needed this is usually obtained by distillation of sodium metal. The appropriate volume of ammonia is condensed as above, and small pieces of clean sodium are added to produce a blue solution. The dry ammonia is then allowed to distill over into a second flask fitted with a dry ice/acetone condenser (the ammonia in the distillation flask must remain blue throughout). The distillation flask is then disconnected from the receiver flask and the reaction carried out as normal.

The work-up for reactions involving liquid ammonia is usually simple. Solid ammonium chloride is added carefully, and the ammonia is allowed to evaporate. The product may then be isolated and purified in the usual way.

14.7   Microwave reactions

Caution: Microwave irradiation is hazardous. Always follow the manufacturer's instructions when using microwave apparatus.

Microwave heating is commonly used in synthetic organic chemistry, and a range of commercial microwave reactors are available for this purpose. The most common reaction setups used for a microwave reaction are shown in Figure 14.7. On a small scale (reaction volumes 0.1–5 cm³), small glass microwave tubes (Figure 14.7a) are typically used. These can be sealed with a pressure cap to allow heating above the boiling point of the reaction solvent. Most commercial microwave reactors are 300 W and

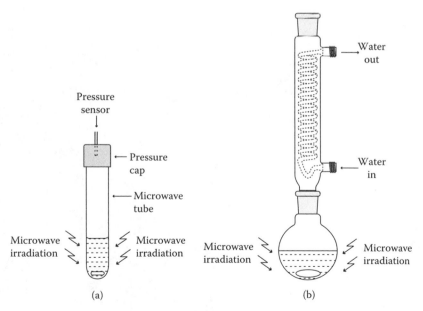

Figure 14.7 Common microwave reaction setups.

can be programmed to run the reaction at a specific temperature (typically up to 300°C) or pressure (typically up to 20 bar) and will adjust the microwave power automatically to achieve this. However, always check the specifications of a particular instrument before starting a reaction to ensure that you are operating within safe limits. Larger scale microwave tubes (up to 50 cm^3) are also available for some commercial apparatus, and most are also capable of heating a conventional open reaction setup involving a round-bottomed flask attached to a condenser (Figure 14.7b).

The main advantage of microwave reactors is that they allow the reaction mixture to be heated and cooled rapidly. This can often result in cleaner reactions than conventional heating. They generally work best when polar reaction solvents are used. An interesting discussion on the pros and cons of microwave heating has recently been published.[3]

References

1. B.D.A. Hook, W. Dohle, P.R. Hirst, M. Pickworth, M.B. Berry, and K.I. Booker-Milburn, *J. Org. Chem.*, 2005, **70**, 7558.
2. T. Veysoglu, L.A. Mitscher, and J.K. Swayze, *Synthesis*, 1980, **12**, 807.
3. J.D. Moseley and C.O. Kappe, *Green Chem.*, 2011, **13**, 791.

chapter fifteen

Characterization

15.1 Introduction

This chapter deals with the type of physical data that are required for the proper characterization of the purified product. No theory is discussed as this is well covered in other sources and, given good data, it is often possible to find a colleague (for example) who will help out if you are unable to interpret a spectrum. With inadequate data it will be difficult to be certain of the structure and purity of your product, and it will certainly be more difficult to interest the colleague referred to previously.

It is important to acquire as much information as possible on your product. It might be "obvious" from the ^{1}H NMR spectrum that the structure is what you think that it should be, but it is still necessary to record (at the very least) the ^{13}C NMR, IR, and mass spectrum. These might simply confirm the ^{1}H NMR data, or they might raise other structural possibilities. The collection of physical data has already been touched upon in Section 3.3, and it is advisable to consult this in addition to the current chapter.

The full set of routine physical data that should be obtained on a pure compound is as follows: IR, UV, NMR (^{1}H, ^{13}C, and any other relevant nuclei), low- and high-resolution mass spectra, m.p. or b.p., and microanalysis (for a new compound). If the compound is optically active, then the optical rotation must also be measured. Only MS and microanalysis from this list are destructive techniques, but modern techniques mean that only a small amount of material need be "sacrificed" for these purposes. Some general points concerning these techniques and the sample requirements are given below.

15.2 NMR spectra

You should always record NMR spectra of organic reaction products. It is normal to record one-dimensional ^{1}H and ^{13}C NMR spectra, along with sufficient additional NMR experiments to allow full assignment of these spectra. The most common NMR experiments that are used to aid assignment of ^{1}H and ^{13}C NMR spectra of organic compounds are listed in Table 15.1.

In addition, nOe experiments are often used to assist structure identification. These experiments help identify protons that are close together in space and can be extremely useful in distinguishing between

Table 15.1 Commonly Used NMR Experiments

NMR experiment	Information obtained
¹H COSY	Identifies proton–proton coupling
45°, 90°, 135° DEPT (¹³C)	Distinguishes between C, CH, CH₂, and CH₃ carbons
¹H/¹³C HSQC	Identifies which proton is attached to which carbon
¹H/¹³C HMBC	Identifies longer range (2-4 bond) proton–carbon connections

stereoisomers. A skilled organic chemist will need to be familiar with the use (and limitations) of all these techniques.

The one-dimensional ¹H NMR spectrum is often the first measurement taken on a reaction product. The size of sample required depends on the type of spectrometer used. A continuous wave machine operating at 90 MHz will need at least 10 mg of a normal organic compound, probably more, but such machines are rare today. Pulsed Fourier transform spectrometers require less material, 5 mg being a normal amount, but high-quality spectra can be obtained using much smaller quantities. Most spectra are measured in deuteriochloroform (CDCl₃), although other solvents will be required from time to time. A typical solvent volume would be approximately 0.4–0.5 cm³ in a 5 mm NMR tube. Routine measurement of the ¹³C NMR spectrum usually requires more sample (25–50 mg), but good spectra can be obtained on less sample; it simply takes more time.

Choice of solvent for NMR spectra is important. Relative chemical shifts are solvent dependent, and ideally, all spectra should be measured in a standard solvent. CDCl₃ is the generally accepted "standard" solvent, and it is advisable to use this where possible. If your compound is not sufficiently soluble in CDCl₃, then an alternative must be found. Lack of solubility in CDCl₃ is usually due to the polarity of your compound; very polar materials and those with extensive, strong, hydrogen bonding networks often need solvents other than CDCl₃. Such materials usually dissolve sufficiently well in hexadeuteriodimethylsulfoxide (DMSO-d₆), tetradeuteriomethanol (MeOH-d₄), or deuterium oxide (D₂O). All these solvents will absorb moisture readily from the air, and so it is advisable to purchase small ampoules and protect them from air once opened. In this way, if the "water" signal should become too great, then a new ampoule can be opened and relatively little solvent wasted.

Which solvent to choose depends on several factors. The most obvious of these is solubility; sufficient compound must dissolve to provide a noise-free spectrum. Given this, the nature of your compound and what you need from the spectrum will be important. Both D_2O and MeOH-d_4 will "exchange out" exchangeable protons such as hydroxyl, amine, or amide protons. If you wish to observe the positions of such protons, it is necessary to use an aprotic NMR solvent such as DMSO-d_6. It is also wise to check in the literature for similar types of compounds and see which solvent has been used in the past for recording their NMR spectra.

If you are seeking to make comparisons of the NMR spectrum of your compound with a known compound, or class of compounds, then it is imperative that you use the same NMR solvent. This also applies to solvent mixtures. Sometimes it is suggested that a small quantity of DMSO-d_6 or MeOH-d_4 be added to $CDCl_3$ if your compound is not sufficiently soluble in $CDCl_3$ alone. This is inadvisable if you wish to make comparisons, unless you are careful to note the exact ratio of the two solvents. Addition of DMSO-d_6 to $CDCl_3$ results in a change of chemical shift of the resonance for residual $CHCl_3$, a peak that is often used as a reference point in ^1H NMR spectra. The chemical shift of the proton of $CHCl_3$ in $CDCl_3$ is 7.27 ppm, in MeOH-d_4 is 7.88 ppm, and in DMSO-d_6 is 8.35 ppm. So be wary of referencing when using mixtures of solvents.

If a polar solvent is needed for NMR spectroscopy, then consider also how you will recover the sample if you need to, as DMSO-d_6 is rather difficult to remove without extensive exposure to high vacuum (but prolonged exposure will usually succeed). In this case, MeOH-d_4 might be preferable, but make sure that you exchange back any "exchangeable" protons by dissolving the sample in MeOH and concentrating under reduced pressure several times.

For high-quality ^1H NMR spectra to be obtained, the solution used needs to be free of paramagnetic metal ions (it usually is) and particles (it usually is not). Place sufficient sample into a clean, small vial (weigh it in if you are unsure how much to use) and add the solvent (ca. 0.5 cm^3) to dissolve the sample. After taking up this solution into a Pasteur pipette, filter it by passing through a small plug of cotton wool in a Pasteur pipette, directly into the NMR tube. This will usually remove small particles and allow a good spectrum to be obtained. Be very careful that no fragments of glass are broken off the pipette during this filtration and on removal of the pipette being used as a filter.

A high-field ^1H NMR spectrum will show up any impurities containing protons. If the compound has been purified, the most common impurity peaks observed in the ^1H NMR spectrum are those from the last solvent used (e.g., solvents used in crystallization or chromatography). This needs to be avoided. It should always be possible to remove trace solvents from your sample unless the b.p. of your product is close to the solvent (which it should not be) or your product is a crystalline solvate (not that common). If residual solvent is present, the sample has not been purified properly. Exposing the sample to high vacuum should remove the last traces of solvent, but sometimes very viscous oils and gums will "hold on" to solvent due to the very slow rate of diffusion. In such instances gently warming the sample under high vacuum usually helps. If this fails, then dissolve the sample in a small amount of $CDCl_3$ and evaporate. Repeat once or twice more and most of the residual solvent should then be $CDCl_3$ rather than (say) ethyl acetate. This will improve the quality of the ^1H NMR spectrum but do not forget that your sample will still be impure on evaporation as it will now contain residual $CDCl_3$.

15.3   IR spectra

The sample again needs to be free of impurities and solvents for IR spectroscopy. There are various methods for sample preparation and which you choose will depend largely on the type of compound and the apparatus available. The amount required for IR spectroscopy is no more than a few milligrams. For a liquid sample the spectrum can be obtained using a thin film obtained by compressing a small drop between sodium chloride plates, or as a solution (usually in chloroform) using a solution cell. The spectra of solids can be recorded either as mulls with a hydrocarbon (e.g., "Nujol") or by mixing with potassium bromide and compressing to form a thin disk or by using an attenuated total reflectance attachment. Which you use will depend upon the facilities available and often on the usual working practice of your department or group.

15.4   UV spectroscopy

UV is only of use if your compound has a characteristic chromophore. There is little point trying to measure weak bands that will provide no information. However, it is of considerable value in several areas of research, for example, natural product isolation, heteroaromatic chemistry, porphyrin and related chemistry, and in the study of dyestuffs. It is also essential if you plan to monitor or purify your compound using an HPLC apparatus equipped with a UV detector.

The amount of material required is usually very small (fractions of a milligram) since the extinction coefficients of useful bands are usually large. The sample must be as pure as possible and is dissolved in the solvent of choice (usually spectroscopically pure ethanol). The concentration must be known accurately so that the extinction coefficients can be calculated and will vary depending upon the type of chromophore. An estimate of the concentration to use can be made if the extinction coefficients for compounds similar to that being studied are available. If these data are not available, make up a solution accurately and dilute it (accurately!) until a reasonable spectrum is obtained.

15.5   Mass spectrometry

There are three pieces of useful information that can be obtained from MS: the molecular mass, composition, and fragmentation pattern of your compound. The accurate molecular mass is of primary importance since this will confirm the elemental composition of your compound. Fragmentation information might be of value for supporting the proposed structure, possibly by comparison with known compounds. The amount required is minimal (less than a milligram), but the material should be reasonably pure. If you are unable to obtain good microanalytical data, the accurate mass measurement in conjunction with good ^{1}H and ^{13}C NMR spectra may provide an acceptable alternative.

15.6   Melting point (m.p.) and boiling point (b.p.)

These are usually straightforward. There are various forms of m.p. apparatus in widespread use, so check carefully on the procedure appropriate for the apparatus available to you. Always obtain a "rough" m.p. before attempting to make accurate measurements, and it is often useful to calibrate the apparatus by measuring a known (pure!) compound with a similar m.p. to the product. If there is a significant discrepancy, then a more reliable apparatus must be used. Do not forget to get the inaccurate apparatus repaired and discard it if necessary The compound needs to be pure and free from dust, and the temperature must be raised very slowly as you near the m.p. If there is a range over which the compound melts (there usually is) then record it; do not estimate an "average" reading. If a capillary tube is used, it is sometimes useful to examine the upper part of the tube for sublimate or distilled decomposition products.

If you have distilled your product to isolate and purify it, then you should already have the information required for reporting the b.p. It is important to quote the range of temperature (if observed) over which the compound distills, the pressure (measured as it is distilling), the vapor temperature (if measured), and the bath temperature. All these will be useful when you or anyone else comes to repeat the work, and most of this information will be required at some time for a publication, report, or thesis.

15.7   Optical rotation

If your product is, or should be, optically active, then the specific rotation ([α]) will need to be measured and recorded. The precise value of this property is dependent on the wavelength of the light used, solvent, concentration, and temperature. Moreover, great care should be taken to exclude any optically active impurities, even if these are present in small quantities, as they might have very large rotations and make your

measurements quite misleading. Clearly it is important to be sure of the purity of your product and to make up the solution carefully and accurately. If you are unsure about making the measurement, make a measurement using a known compound before you try to measure the rotation of your product (assuming that the specific rotation of your product is not known). Usually you will use the sodium-D line and measure at ambient temperature, but be sure to record the concentration, solvent, wavelength, and temperature along with the actual value of the measured specific rotation. Occasionally optical rotatory dispersion and/or circular dichroism spectra will be required. These measurements will usually be made by specialists, and the specific requirements for your particular type of compound are best discussed with them.

15.8   *Microanalysis*

Ideally all new compounds and known compounds that have been prepared by a new route should have their purity assessed by microanalysis. It goes without saying that the sample must be homogeneous and free from dust and other impurities. For solids, careful recrystallization using pure, filtered solvents followed by equally careful isolation by filtration and drying will usually suffice. Ideally, for oils, distillation followed by sealing in an ampoule usually will provide acceptable samples. However, provided that distilled solvents are used, care is taken in fraction collection to avoid particles of silica in the fractions, and great attention is applied to solvent removal, it is often possible to obtain acceptable microanalysis on samples obtained by flash column chromatography. Solids must be dried in high vacuum in a drying pistol to remove traces of solvent and submitted in clean, dry vials.

Microanalytical data should be within 0.3–0.5% of theory. If this is not the case, then the sample is either impure or not what you think it is. Try to fit the data to a sensible molecular formula, and if a good fit is found, then try to work out a structure that might correspond to the formula and that would also be compatible with the rest of the analytical data (NMR, IR, mass spectra, etc.). Occasionally compounds will crystallize with a molecule of solvent (or water). Try molecular formulae that include a molecule of the solvent of crystallization (or possibly water). If this works, then you should be able to detect this molecule of crystallization in the ^1H or ^{13}C NMR. Occasionally, samples will absorb water from the air, and if you and the analyst are unaware of this, inaccurate results can be obtained. This hygroscopic behavior is easily tested by carefully weighing a freshly purified sample and checking if it gains weight after exposure to the atmosphere for a few hours. An IR spectrum of the sample before and after weighing should reveal any serious contamination with water. This problem can usually be solved by careful sample preparation, isolation, and sealing of the sample in a vial or ampoule. Make sure that you provide the analyst with the information regarding the hygroscopic nature of the material.

If you cannot obtain a fit with any reasonable molecular formula, then accept that the sample must have been less pure than you thought. Repurify and try again.

15.9   Keeping the data

If you work for any length of time in the laboratory you will soon accumulate a large quantity of analytical data. It is important that you keep these safe and in proper order, using an unambiguous cross-referencing system so that you (and anyone else) can locate the spectra or measurements that apply to the product of a particular experiment. See Chapter 3 for detailed advice on how to do this. Spectra are best kept in clearly labeled folders or binders of some description, preferably ones that allow for removable attachment of the spectra. Other data should be recorded in the laboratory notebook along with the experimental write-up. If data sheets are used, then all the data should be recorded on these as they are measured.

chapter sixteen

Troubleshooting
What to do when things don't work

Do not despair, yet. Some reactions will not work, for proper chemical reasons associated with the substrate. These may or may not be "obvious" (many things only become "obvious" with hindsight). However, do not immediately jump to the conclusion that your desired reaction will not work. Before you can conclude that this is the case, a number of possibilities must be explored.

The first and most obvious is to make sure that all the starting materials, solvents, and reagents are pure, dry, and free from other solvents. Also, check the starting material carefully to ensure that it is indeed what you think it is. A critical perusal of *all* the analytical and spectroscopic data will usually be sufficient. If you have prepared several batches of starting material, then make sure that the spectra that you check are from the batch that you used in the failed reaction. It is also worth rerunning the ^{1}H NMR spectrum of your starting material to ensure that it has not partially decomposed or absorbed moisture.

If the starting material was purchased from a chemical supplier, check that it really is what it purports to be by recording and checking the ^{1}H NMR and IR spectra and any other data that might be available. If a reagent has been purchased, then it is usually possible to check its reactivity (and your technique) by carrying out a known, reliable, literature reaction. If it is possible to purify the reagent, do so, but make sure that you are handling it correctly (Chapter 6). It will not be practical to purify some reagents, for example, alkyllithiums, and in this case, the quality should be checked by titration where possible. It is unwise to purchase a reagent (from any source) and to take for granted the quoted molarity and purity; even the most reputable suppliers are fallible and occasionally make mistakes.

Once you are certain that the problem does not lie with the starting material or reagents, check the solvent. Many reactions will not work if the solvent is not anhydrous, and methods for obtaining anhydrous solvents are provided in Chapter 5. THF is very commonly used and usually dried by distillation of sodium/benzophenone. However, it is possible to collect *wet* THF from a bright blue distillation pot (which only indicates that the solvent is dry in that part of the still). This problem is often encountered when a still head is used to collect the solvent, but insufficient time has been allowed at reflux before collection is commenced. One way to check a sample of the THF to be used (before introducing it into the reaction flask) is to add a little sodium hydride (dispersion in oil) (*care*). If immediate hydrogen evolution is observed, the solvent is wet. The remedy is to allow more time at reflux before collecting the solvent; if this does not work, the drying agents might need recharging.

Another source of water could be the inert gas supply that you are using. Make sure that the drying agent (if used) used is still working and renew it if necessary. Failing this, try the reaction under argon, which usually contains much less moisture than nitrogen. Check the manifold and renew any suspect tubing.

If the starting material remains unreacted after the attempted reaction, and if the starting material, reagents, solvent, and inert gas are all as they should be, consider the possibility that the reaction might be reversible. It might be that you are actually under equilibrating conditions and that the reaction is under "thermodynamic control" with the equilibrium favoring the starting materials. Conditions and reagents might need to be changed to avoid this; you might require conditions of "kinetic control." Only then consider that the problem might be you! It is possible that you might be inadvertently carrying out the reaction in such a way that it will not work. For example, is the temperature correct? Are the *concentrations* of reactant and reagent correct? Is too much or too little time being allowed at a particular stage? There are many possibilities. To test this, it is advisable to carry out the same reaction but use a substrate that is known from the literature to react properly. If this is successful and your desired reaction is not, then you might have found a reaction that does not work on your substrate, and alternative conditions (different metal ions, different Lewis acid, etc.) might be required. Above all, if the reaction is an important one, do not give up; perhaps you could take the opportunity to develop a new reagent or procedure that will work.

If the starting material decomposes under the reaction conditions, then consider carefully any possible "chemical" reasons that might be responsible. For example, basic conditions might be causing deprotonation and/or elimination at an undesired site or sites in the compound, and acidic conditions might be responsible for an unexpected cyclization or protecting group loss. There are often several possible sources of instability toward a particular set of reaction conditions, and it will be necessary to examine all possibilities.

It is likely that the reaction will be closely related to a known process, and if this is so, then be critical in your comparison of reaction conditions and reactants. For example, if an enolate alkylation is being attempted, but appears not to work at a temperature below which the enolate decomposes, consider the reactivity of the alkylating reagent. Is it really as reactive as in the example that you are following? If not, then the appropriate change will need to be made.

Decomposition of your starting material might be occurring at several stages in the reaction. This can often be ascertained by careful TLC of the reaction mixture at various stages. For example, if you are attempting enolate alkylation, decomposition might be occurring on addition of base, before addition of the alkylating agent. Analysis by TLC after the addition of base (before addition of alkylating agent) will show if decomposition is occurring. If all is well, TLC after addition of the alkylating agent will also indicate if decomposition is occurring later in the procedure. If all looks fine and a product appears to have been formed (by direct TLC of the reaction mixture), then the product might be decomposing on work-up. This is easily checked by quenching samples of the reaction mixture into solutions of varying pH and buffer solutions, followed by TLC analysis.

If all else fails, and you are attempting to follow a published procedure, or to apply a published procedure to a new system, then it might be worth contacting the senior author on the publication that you are following. Many such authors are willing to provide further information and advice. If such help is forthcoming and allows you to succeed, then be sure to let the supplier of such information know and send a copy of your manuscript, should you wish to publish your results. Be sure to acknowledge this help in such publications. This simple courtesy will avoid any misunderstandings and ensure good relations in the future.

The single most important piece of advice in troubleshooting a reaction is to be certain that you identify the problem *before you try to remedy it.* This might seem to be obvious, but it is surprising how often this advice is not followed. Try hard to keep an open mind, to analyze the problem critically and logically, and to gather evidence to support your identification of the problem. When you have a clear idea what the problem actually is, you can begin to think of potential solutions that are likely to have the desired effect.

chapter seventeen

The chemical literature

17.1   Structure of the chemical literature

Consulting the literature is an essential element of chemical research. Whether you want to confirm the identity of your latest product or check the feasibility of an exciting new idea, it would be both unscientific and counterproductive not to conduct a thorough literature search. Moreover, it is vital to the success of your work to keep abreast of developments in organic chemistry in general and in your area in particular. The problem is that finding chemical information and keeping in touch with current developments are difficult and time-consuming tasks.

We are the beneficiaries of almost one and a half centuries of research in organic chemistry. The accumulated output of that effort is an enormous body of data collected in a vast literature. It is estimated that well over half a million articles (papers, patents, books, etc.) are published each year and the volume of publications will probably continue to rise. Searching such a huge body of work is a formidable problem, but it must be emphasized that the time spent reading the literature is often more than repaid by the experimental time saved as a result.

This chapter is intended as a practical guide to efficient searching of the chemical literature; the main part is devoted to a description of the most important access routes to the primary literature and a discussion of methods of tackling some common types of literature search. The chapter concludes with a section on methods of keeping in touch with the current literature. The reader is referred to a number of other texts for more detailed information.[1-4]

Almost all chemical information is originally published in research journals, in patents, and in theses. These sources are called the primary literature, and the goal of most literature searches is to find the original reports containing the required information. There are thousands of journals that publish papers on chemistry, but in practice, the great majority of papers that are of interest to the organic research chemist appear in just a hundred or so of these. This is still a dauntingly large body of information, but there are several routes by which it can be searched and specific items of information located.

An important route, and one that is rapid and easy, is to tap the chemical knowledge of your colleagues and supervisors. Many of the people working around you are likely to be experts in their own fields. Another route is to use the secondary literature. This comprises review articles and books, in which the original literature has been organized and summarized, and reference books in which particular kinds of data have been collected together. Of course, finding the appropriate review or handbook is a problem in itself. A third route is via paper indexes and electronic databases that can be searched for literature references to information on a given compound, or procedure, or author, and so on. Prominent among the printed indexes was Chemical Abstracts (CA), which covered over 12,000 different sources and contains short summaries of just about every paper published on a chemical topic, as well as comprehensive indexes to these abstracts, and hence to the original papers. This index ceased publication in 2009, but copies are still available in many libraries, and this is a useful resource for searching the chemical literature up to 2009. CA has been superseded by the electronic database SciFinder.

To aid discussion of these information sources, we have split them into two categories: paper-based information sources (journals, books, series, etc.) and electronic sources (computer databases), because although the type of information available from each source may be the same, the coverage and methods for searching them often differ significantly.

17.2   Some important paper-based sources of chemical information

This section contains a description of structure, strengths, and weaknesses of three of the most important paper-based tools for locating information in the primary literature: CA, *Beilstein*, and the Science Citation Index (SCI). It is followed by a complementary section on how to carry out some specific kinds of searches. All three of these paper-based information sources have now been superseded by electronic databases and these are discussed later in Section 17.3.

17.2.1   Chemical Abstracts

CA ceased publication in 2009 and its content is now incorporated in the electronic database SciFinder (Section 17.3.1). It consists of two main parts: abstracts of every paper containing new chemical information and indexes that provide access to the abstracts and thence to the original literature. It was published weekly and each issue contained a keyword index and an author index. The weekly issues are collected in volumes covering a 6-month period (1 year, prior to 1962), and each volume contains author, chemical substance, formula, general subject, patent, and ring system indexes. Every 5 years (10 years, prior to 1957), the indexes for the 10 volumes are combined to give Collective Indexes.

A search of CA should begin with the appropriate index of the most recent volume available and should progress backwards through the other volumes until the beginning of the period covered by the most recent Collective Index, at which stage the Collective Indexes are used to search the literature back to 1907. The most useful indexes are the chemical substance, formula, and general subject indexes. Consulting the indexes will, in the first instance, lead to references to the abstracts, not directly to the literature. The references to the abstracts take the following form: **90**:*108753h*, where **90** is the CA volume number and the abstract number is *108753*. Since 1967, abstracts have been numbered sequentially in each volume. Before that, the references were of the form **46**:*13761a*, where **46** is the volume number, *13761* is the column number, and the letter *a* indicates that the abstract is at the top of the column. Earlier still, a numerical superscript was used to indicate the position of the abstract in the column. The letters *B*, *P*, or *R* before an abstract number indicate that the original work is a book, patent, or review respectively.

The abstracts contain full bibliographic details of the original paper and a summary of the principal new findings reported in the paper. A glance at the abstract will tell you if the original journal is likely to be accessible, what language the paper is in, and most importantly it will give an indication of whether the paper really does contain the information you require. Remember that many of the compounds described in the original paper will not be mentioned in the abstract but will be contained in the indexes. If the abstract looks promising, all that remains is to locate the journal and consult the paper.

An *Index Guide* was published every 18 months and contains invaluable information on the use of the indexes and the system of nomenclature used in CA. It is essential reading for serious users of CA. Finally, the *Ring Systems Handbook* and its predecessors, the *Ring Index* and the *Parent Compound Handbook*, contain information on ring and cage systems and gives the names under which ring systems can be found in the Chemical Substance Index.

The great strengths of CA are that it provides comprehensive literature coverage, and it has extensive indexes. However, the coverage of the literature in the early years was not so rigorous, and *Beilstein* (see Section 17.2.2) provides more thorough coverage of the pre-1949 literature.

17.2.2   Beilstein

Beilstein's Handbuch der Organischen Chemie, or *Beilstein* for short, is a huge (>500 volumes published between 1918 and 1998) reference work that contains physical and chemical data for over one and a half million compounds. It has now been superseded by the electronic database Reaxys (Section 17.3.2). The compounds are organized according to a unique classification system, and each volume contains a subject (actually compound) index and a formula index. Comprehensive literature coverage is attempted, so *Beilstein* contains essentially all the compounds of a given class that were prepared from the beginning of organic chemistry to the date of publication of the most recent volume covering that class of compounds. A considerable amount of critically reviewed information, with references to the primary literature, is provided for each compound. This data includes the molecular structure, natural occurrence, methods of preparation, physical properties including references to papers containing spectral data, and chemical properties.

The handbook consists of the original series of 29 volumes (the *Hauptwerk, H*), covers the literature up to 1909, and is divided as follows: volumes 1–4, acyclic compounds; volumes 5–16, carbocyclic compounds; volumes 17–27, heterocyclic compounds; volume 28, General Subject Index; and volume 29, General Formula Index. In addition, there are four complete supplementary series (*Erganzungswerk, EI, EII*, etc.) covering the literature up to 1959. The fifth supplement is not comprehensive but covers selected heterocyclic compounds up to 1979. Cumulative indexes are only available for the pre-1930 literature. It is relatively easy to find data for any compound reported before 1930 by checking the cumulative formula index (Volume 29 in three subvolumes), which will give the volume and page numbers for the entries in H, EI, and EII. Compounds of the same class will appear in the same volume of each series (although some of the volumes are now divided into several subvolumes). Thus, if a compound is located using the cumulative index, it is a simple matter to locate references in the later series, using the volume indexes for the same volume or using the *Beilstein System Number*. If a compound is not contained in the cumulative index, it is best to try to identify which volume it should be contained in, using the classification system. A booklet explaining how to use the *Beilstein* system may be available from your librarian.[5]

Beilstein is the best and most comprehensive source of data for organic compounds prepared before 1930 and it is not particularly difficult to use. Its major weakness is the lack of data and cumulative indexes for more modern work.

17.2.3 Science Citation Index (paper copy)

The SCI is now incorporated into the Web of Science electronic database (Section 17.3.3) but was originally published in paper form. The SCI was published every 2 months and was cumulated annually. It includes coverage of all the major chemistry journals. It is a combination of three indexes that provided coverage of all the important publications in the physical sciences:

1. The *Source Index* lists the bibliographic details for the publications for each author/organization.
2. The *Permuterm Index* is based on combinations of keywords in the titles of the articles published in the journals that are covered by the SCI. For example, under the keyword "epoxide" will be a list of other keywords such as "stereoselective," which occur in association with "epoxide." For each pairing, there is a list of authors' names, and looking up these in the *Source Index* will lead to the references for the original work.
3. The *Citation Index* allows you to search the literature *forward* in time. The index entries are the names of the first authors of each paper that was cited in any paper published during the period covered by that issue. Its use is best illustrated with the aid of an example. Suppose that a researcher found a paper published in 1980, by S. Smith et al., which contained some very interesting results, and wanted to know if any further relevant work had appeared. The researcher would consult the annual indexes of the SCI from 1980 onwards. Each index contains an alphabetic list of *first* author's names and under S. Smith's name would be a chronological list of his/her publications. Under the entry for the paper of interest to our researcher, there would be a list of papers that cited it, published during the period covered by that index. It is reasonable to assume that any workers who followed up on the results in the Smith paper would have cited it in their own publications. Thus, the list of papers that cited the original should include most of the work subsequently carried out in that area. Hence, if you find an important paper, you can use the SCI to get a list of all the papers that referred to it subsequently. The drawback is that many of the references you find will not be relevant to your interest and there is no way of knowing which are relevant except by consulting the *Source Index*, which gives the titles of the papers, or by consulting the papers themselves.

The *Citation Index* is an extremely useful tool, but as the paper copy is no longer published, it will not allow you to search forward in time to the current date. For this reason, it is only worth consulting if you do not have access to an up-to-date electronic database containing the same information.

17.3 Some important electronic-based sources of chemical information

The development of the Internet and mobile devices has brought about a revolution in the way in which we access, search, and store chemical information. The major chemical indexes and databases are now only produced in electronic form. Although most of the primary literature is still available in paper form, almost all are also published online and some chemistry journals are exclusively published in digital form. In the near future, it is likely that all paper copies of journals will cease and the entire scientific literature will only be available in digital form.

The advantages of so-called online access include much greater speed of access and the ability to search for information in a wide variety of ways. There is far more information available online than could possibly be stored in most institutional libraries and wireless networking means that this information can be accessed from almost anywhere. In addition, some electronic databases include material that cannot otherwise be searched directly, for example, the full text of a journal or reference book.

The principal advantage of an electronic database over the paper version is much greater flexibility and power in carrying out searches. For example, it is possible to combine a number of keyword searches in one using logical operators (e.g., oxidation AND [alcohol OR aldehyde]). Even more importantly, it is possible to search structures or substructures via graphical input, searches that were almost impossible using printed indexes.

There are disadvantages too. None of the electronic databases is comprehensive and all contain some errors in the data. If you use these as your sole means of searching the literature, it is likely that you could miss important information. Another problem is that the software for online searching can be relatively complex and hence more difficult to learn. As a result, online searches by inexperienced users can be unreliable. The use of graphical input and output makes database searching much easier, but a good understanding of the software is still essential. Finally, the cost of hardware, software, consumables, and the searches themselves can be considerable.

In the following, we briefly outline details of some of the more important chemical databases that are currently available. All these databases are commercial and require registration and payment of an annual fee to use. Access to such databases will vary from institute to institute, so you should ask at your library for a list of those available to you. It is not possible to describe the operation of all the chemistry databases here, but you should be able to get help from your librarian or your supervisor and learn how these very powerful tools can help you.

17.3.1 SciFinder

This is the electronic successor to *CA* and is one of the largest databases available that relates to synthetic organic chemistry. SciFinder includes four key databases:

1. CAplus. This contains key information from the chemistry literature published between 1907 and the present. It also contains selected coverage of material published between 1808 and 1906.
2. CAS REGISTRY. This contains specific information about chemical substances published from 1957 to the present. It also includes some classes of compounds back to the early 1900s.
3. CASREACT. This is a database containing information about chemical reactions and covers the literature back to 1907.
4. MEDLINE. This database covers the biomedical literature back to 1950.

SciFinder is updated weekly and covers all the key organic chemistry journals and patents. It allows the user to search these databases using a range of keyword and graphical search facilities. It also provides information on commercially available materials and regulatory data. This is an essential resource for full-time organic chemists.

17.3.2 Reaxys

Reaxys is another huge chemical database and contains selected data back to 1771. Its content is derived partly from *Beilstein*, but also contains information from a number of other key sources and is continually being updated.

Reaxys allows a variety of graphical searches to be performed (e.g., structure, substructure, and transformation). It also has methods for filtering the results that are particularly useful for organic chemists (e.g., reactions can be filtered by the reagent or solvent used). It also provides information on commercially available materials.

The functionality of Reaxys and SciFinder are similar and appear to be converging, but the content differs and neither is comprehensive. Consequently, both are essential resources for synthetic organic chemists. Assuming you have a subscription, you should always check both when searching for chemical information.

17.3.3 Web of Science and SCOPUS

This Web of Science is the electronic version of the SCI and SCOPUS is a closely related product. Both can be searched by a variety of text inputs including keywords, author's name, institute name, and citation. They both have a range of tools that allows analysis of citation data, enabling you to search a subject or author both backwards and forwards in time. Both databases are regularly updated and are essential tools for the synthetic organic chemist.

17.3.4   Cambridge Structural Database (CSD)

This database contains bibliographic and three-dimensional structural results from crystallographic analyses of small organic and organometallic compounds. Both x-ray and neutron diffraction studies are included. The database covers structures up to about 500 atoms (including hydrogens) and as such is an invaluable source of structural information. Currently, there are over 500,000 entries, which are increasing at the rate of about 37,000 a year. It can be searched by both keyword and graphical input.

17.3.5   The World Wide Web

Although not strictly a database, the World Wide Web (WWW) is an almost limitless source of all types of information, including chemical information. Much of the chemical content on the WWW is freely accessible and can be searched using the usual Web search engines. When using chemical information from the WWW, it is important to bear in mind that the content of most sites is not regulated or peer reviewed. So *always* check the primary literature sources to make sure that the key information is correct before attempting a reaction based on information sourced in this way.

17.4   How to find chemical information

17.4.1   How to do searches

The following are some basic rules for guidance in searching the literature:

1. Clearly define the goals of your search.
2. Discuss the problem with your colleagues and supervisors; they may have some valuable expertise.
3. Determine which information sources are available to you and which to use.
4. Start with the current literature and work backwards; the recent literature will contain references to earlier work.
5. When you find a key paper, check it carefully for references to relevant earlier work and also work forward in time by carrying out a citation search.

6. Keep a complete record of your search, noting all the sources you used and the information you obtained. This is invaluable if you have to carry out related searches later. It is advisable to keep a separate notebook for recording your literature searches; the information you accumulate will build into a very useful resource. If you perform electronic searches, it is usually possible to download the search results so that you can refer to them again later if necessary.

7. Sign up for "chemistry journal alerts" in key areas of interest. Most online journals will allow you to set up e-mail alerts that will notify you whenever a new paper in your area appears. They will also notify you whenever a new issue of the journal is published.

17.4.2   How to find information on specific compounds

The two main databases that allow you to search for information on particular compounds are SciFinder and Reaxys. Information on relatively simple compounds can also often be obtained from handbooks that may be available in your laboratory, such as the following:

1. The catalogues of major chemicals suppliers.
2. CRC Handbook of Chemistry and Physics, CRC Press.

17.4.3   How to find information on classes of compounds

Finding information about a broader area, such as a class of compounds, is usually more difficult than finding data about a specific compound. It is usually best to begin by consulting books on the area and then progress to more specialized monographs and reviews, before consulting the primary literature.

Good starting places for the older literature include *Comprehensive Organic Chemistry, Comprehensive Organometallic Chemistry,* and *Comprehensive Heterocyclic Chemistry* (Pergamon Press), which are multivolume texts giving a detailed overview of the title areas. A much more detailed treatment of many common classes of compounds is contained in the series *The Chemistry of the Functional Groups* edited by S. Patai (Wiley). This excellent series contains reviews on all aspects of the chemistry of one particular functional group. Other multivolume series include *The Chemistry of Heterocyclic Compounds: A Series of Monographs* (Wiley) and *Rodd's Chemistry of Carbon Compounds* (Elsevier). In addition, there are many review series devoted to the chemistry of particular classes of compounds, including the *Specialist Periodical Reports* published by the Royal Society of Chemistry.

Finding individual books or reviews is more difficult. A good starting point is your library; glance along the shelves and consult the catalogues. You can also search for reviews by keyword using the Web of Science or SCOPUS.

17.4.4   How to find information on synthetic methods

The two main databases that allow you to search for synthetic transformations are SciFinder and Reaxys. There are also several major works devoted specifically to synthetic methods. Most notable are the *Encyclopedia of Reagents for Organic Synthesis* (2005, Wiley), which contains short reviews on almost all the common reagents used in organic synthesis, and *Reagents for Organic Synthesis* (L.F. Fieser and M. Fieser, Wiley), which was published between 1967 and 2005. These two works constitute the best source of information on the preparation, purification, and use of synthetic reagents. The *Encyclopedia of Reagents for Organic Synthesis* is also available online as *e*-EROS. A relatively inexpensive alternative is the single-volume work *Comprehensive Organic Transformations* (R.C. Larock, VCH Publishers), which outlines a large range of functional group transformations and is an excellent starting place when searching for such information.

Other useful texts include *Comprehensive Organic Synthesis* (Pergamon), a nine-volume overview of synthetic methods, *Compendium of Organic Synthetic Methods* (Wiley), *Organic Reactions* (Wiley), *Advanced Organic Chemistry, Reactions, Mechanisms, and Structures* by J. March (Wiley), and *Organic Syntheses* (available free online), a compilation of carefully checked procedures with full experimental details, which is an excellent source of representative synthetic procedures.

17.5   Current awareness

Keeping in touch with the current literature is a difficult and time-consuming exercise, but it is vital to your development as a chemist and to the success of your research. You should aim to read through *at least* 10 of the most important journals in your field, and the best way of doing this is to set aside a specific period each week for reading the latest issues. In addition, you should scan the graphical abstracts of all other chemistry journals relative to your area of research. Signing up for all the relevant "journal content alerts" or RSS feeds will mean that you are notified whenever a new issue appears.

All of this effort will be wasted if you do not keep good records of what you have read. Building your own computer database is a useful way of doing this. Download a PDF copy of each important paper and file it in a way that allows you to retrieve the information rapidly.

References

1. G. Wiggins, *Chemical Information Sources*, McGraw-Hill, New York, 1991.
2. H. Schultz, *From CA to CAS ON-LINE*, VCH Verlagsgesellschaft, Weinheim, 1988.
3. Y. Wolman, *Chemical Information: A Practical Guide to Utilization*, 2nd ed., Wiley, Chichester, 1988.
4. R.E. Maizell, *How to Find Chemical Information*, 2nd ed., Wiley, Chichester, 1987.
5. *How to Use Beilstein*, Beilstein Institute, Springer-Verlag, Berlin, 1978.

Appendix 1: Properties of common solvents

Solvent	Bp (°C)	Mp (°C)	ε	Density	¹H δ (Ref. to TMS) (300 MHz, CDCl3)	¹³C δ (Ref. to TMS) (300 MHz, CDCl₃)	Preliminary drying	Rigorous drying	TLV (ppm)
Acetic acid	118	17	6.2	1.049	2.08 s, 10–13 br s var	20.7, 177.6	Acetic anhydride	Acetic anhydride	10
Acetone	56	−94	20.7	0.790	2.13 s	30.7, 206.5	3A sieves	3A sieves	1000
Acetonitrile	82	−46	36.2	0.777	1.97 s	1.7, 116.2	Potassium carbonate	Phosphorus pentoxide	40
tert-Butanol	82	25	3.5	0.850	1.24 s, 1.35 br s var	31.2, 69.2	Calcium hydride	Calcium hydride	
Chlorobenzene	132	−46	5.6	1.106	7.28 br m	126, 129, 130, 134	Not necessary	Calcium hydride	75
Chloroform	62	63	4.7	1.480	7.24 s	77.0	Alumina	Phosphorus pentoxide	10
Dichloroethane	83	−35	10.4	1.235	3.71 s	44.4	Not necessary	Calcium hydride	10
Dichloromethane	40	−97	8.9	1.325	5.28 s	53.4	Not necessary	Calcium hydride	100
Diethyl ether	35	116	4.3	0.714	1.18 t, 3.45 q	15.2, 65.8	Calcium chloride; Na	Sodium/benzophenone	400
Dimethoxyethane	83	−58		0.850	3.36 s, 3.51 s	59.0, 71.8	Calcium chloride; Na	Sodium/benzophenone	
Dimethylformamide	152	−61	36.7	0.945	2.81 s, 2.89 s, 7.94 s	31.4, 36.4, 162.4	Calcium hydride	Phosphous pentoxide	10
Dimethyl sulfoxide	189	18	49.0	1.101	2.62 s	40.6	Distillation	4A sieves	

Solvent	bp	mp	ε	Density	^{1}H NMR	^{13}C NMR	Drying agent	Better drying agent	Amount
Dioxan	102	12	2.2	1.034	3.66 s	67.0	Calcium chloride: Na	Sodium/benzophenone	50
Ethanol	78	−114	24.3	0.785	1.18 t, 2.05 br s, 3.65 q	18.3, 58.2	Magnesium	3A sieves	1000
Ethyl acetate	77	−84	6.0	0.900	1.22 t, 2.01 s, 4.09 q	14.1, 20.9, 60.3, 171.0	Potassium carbonate	4A sieves	400
HMPA	235	7		1.030	2.59 d	36.8	Calcium hydride	Calcium hydride	
Methanol	64	−97	32.6	0.791	1.94 br s var, 3.42 s	50.5	Magnesium	3A sieves	200
Nitromethane	101	−28	38.6	1.137	4.33 s	62.4	Calcium chloride	4A sieves	100
Pyridine	116	−42	12.3	0.982	7.24 m, 7.63 m, 8.58 m	123.6, 135.8, 149.8	Calcium hydride	Calcium hydride	5
Tetrahydrofuran	66	−108	18.5	0.805	1.82 m, 3.72 m	25.6, 67.9	Calcium chloride; Na	Sodium/benzophenone	200
Toluene	111	−95	2.4	0.867	2.32 s, 7.17 br s	21, 125, 128, 129, 138	Not necessary	Calcium hydride	100
Water	100	0	78.5	1.000	1.56	—	—	—	10

Appendix 2: Properties of common gases

Gas	Mol. wt.	Bp (°C)[a]	Mp (°C)	Density of gas (g/L)[b]	Density of liquid (g/mL)[c]	Main hazards[d]	TLV (ppm)
Acetylene	26.04	−84		1.11		A, F, explosive when pressurized	25
Ammonia	17.03	−33	−78	0.71	0.68	T, C, F	25
Argon	39.94	−189	−186	1.66	1.4	A	
Boron trichloride	117.19	13	−107	4.85		T, C	
Boron trifluoride	67.81	−100	−127	3	1.59	T, C	1
Carbon dioxide	44.01	sublimes	−78	1.83		C	5000
Carbon monoxide	28.01	−191	−207	1.16	0.79	T, F	50
Chlorine	70.91	−34	−101	2.97	1.56	T, C	1
Ethylene	28.05	−104	−170	1.17	0.57	F	
Ethylene oxide	44.05	10.7	−112	1.82	0.88	T, F	5
Fluorine	38.00	−188	−220	1.57	1.5	F, C	1
Helium	4.00	−269	−272	0.17	0.12	A	
Hydrogen	2.02	−253		0.08	0.07	F	
Hydrogen bromide	80.92	−67	−87	3.34	2.16	T, C	3
Hydrogen chloride	36.46	−85	−114	1.52	1.19	T, C	5
Hydrogen fluoride	20.01	20	−83	0.94		T, C	3
Hydrogen sulfide	34.08	−60	−85	1.43	1.0	T, C, F	10
Isobutylene	56.11	−6.9	−140	2.39	0.63	F	
Methanethiol	48.117	6	−121	2.14	0.89	T, F	0.5
Nitric oxide (NO)	30.01	−152	−164	1.24	1.27	T	25

Nitrogen	28.01	−196		1.25	0.8	A
Nitrogen dioxide (NO$_2$)	46.01	21	−9	3.30	1.45	T, C	3
Oxygen	32.00	−183	−218	1.33	1.14	O
Phosgene	98.92	8	−128	4.10	1.41	T, C	0.1
Sulfur dioxide	64.06	−10	−75	2.70	1.46	T, C	2

[a] At 1 atm.
[b] At 20°C and 1 atm.
[c] At Bp.
[d] A = Asphyxiant, C = Corrosive, F = Flammable, O = Oxidizing, T = Toxic.

Appendix 3: Approximate pK_a values for some common reagents versus common bases

Reagent to be deprotonated	pK_a[a]	Bases	pK_a (of protonated base)[a]
ArSH	7	Et$_3$N	11
RCH$_2$NO$_2$	9	Et$_2$NH	11
RCOCH$_2$CN	9	NaOH	16
RCOCH$_2$COR	9	NaOEt	18
RSH	10	KOt-Bu	19
RCOCH$_2$CO$_2$R	11	LiHMDS	26
RCO$_2$CH$_2$CO$_2$R	13	NaH	35
Cyclopentadiene	15	NaNH$_2$	35
RCH$_2$CHO	17	LDA	36
RCOCH$_3$	20	n-BuLi	50
RC≡CH	23	tert-BuLi	>50
RCH$_2$CO$_2$R	25		
RCH$_2$CN	25		
ArH	41		
R$_2$C = CHCH$_3$	44		

Note: Acidic hydrogen is highlighted in bold.

[a] These figures are *only* a rough guide, the actual pK_a values will depend on the nature of R and Ar. A pK_a difference of >4 between base and reagent will give complete deprotonation.

Appendix 4: Common Bronsted acids

Acid	Mol. wt.	pK$_a$ (H$_2$O)	Specific gravity	Weight%	Molarity (M)
Acetic acid (CH$_3$CO$_2$H)	60.1	4.8	1.05	100	17.5
Formic acid (HCO$_2$H)	46.0	3.8	1.22	100	26.5
Hydrofluoric acid (HF)	20.0	3.2	1.17	51	29.0
Phosphoric acid (H$_3$PO$_4$)	980	2.1, 7.2, 12.3	1.70	85	14.7
Nitric acid (HNO$_3$)	63.0	−1.3	1.42	69	15.6
Sulfuric acid (H$_2$SO$_4$)	98.1	−3.0, 2.0	1.84	98	18.0
Hydrochloric acid (HCl)	36.5	−8.0	1.18	38	12.0
Hydrobromic acid (HBr)	80.9	−9.0	1.50	49	9.0
Perchloric acid (HClO$_4$)	100.5	−10	1.53	60	9.1

Appendix 5: Common Lewis acids

Lewis acid	Compatible solvents	Comments
Aluminum trichloride	Hydrocarbons, halogenated	Strong, widely used in Friedel–Crafts
Boron tribromide	Hydrocarbons, halogenated	Strong, used to cleave ethers, acetals
Boron trichloride	Hydrocarbons, halogenated	Strong, used to cleave ethers, acetals
Boron trifluoride etherate	Many solvents	Moderate, very versatile
Diethylaluminum chloride	Hydrocarbons, halogenated	Moderate, useful for proton-sensitive reactions
Ethylaluminum dichloride	Hydrocarbons, halogenated	Moderate, useful for proton-sensitive reactions
Ferric chloride	Hydrocarbons, halogenated	Moderate
Mercuric chloride	Many solvents	Weak, useful for cleaving C-S bonds in thioacetals
Lanthanide shift reagents	Many solvents	Weak, useful for reactions involving sensitive dienes, ethers
Magnesium bromide	Hydrocarbons, halogenated	Moderate
Magnesium chloride diethyl etherate	Hydrocarbons, halogenated, ethers	Moderate
Scandium triflate	Many solvents	Moderate, very versatile, water tolerant
Silver chloride	Many solvents	Moderate

(Continued)

Lewis acid	Compatible solvents	Comments
Silver triflate	Many solvents	Moderate, used to generate carbocations
Tin tetrachloride	Hydrocarbons, halogenated	Strong, very versatile
Titanium tetrachloride	Hydrocarbons, halogenated	Strong, very versatile
Trimethylsilyl iodide	Hydrocarbons, halogenated, MeCN	Strong, used to cleave ethers, acetals, esters
Trimethylsilyl triflate	Hydrocarbons, halogenated	Strong, used with silylated reagents
Zinc bromide	Hydrocarbons, halogenated	Moderate, versatile
Zinc chloride	Hydrocarbons, halogenated, ethers	Moderate, versatile

Appendix 6: Common reducing reagents

Table A6.1 Hydride reducing agents

Reagent	Typical solvents	Typical temperature (°C)	Functional groups reduced
LiBH$_4$	Tetrahydrofuran	0 to RT	Ester → alcohol Ketone → alcohol Aldehyde → alcohol
Li(Et$_3$BH) (Superhydride)	Tetrahydrofuran	−78 to RT	Ester → alcohol Ketone → alcohol Aldehyde → alcohol Alkyl halide → alkane Epoxide → alcohol
Li(s-Bu$_3$BH) (L-Selectride)	Tetrahydrofuran	−78 to RT	Ketone → alcohol Aldehyde → alcohol Alkyl halide → alkane
NaBH$_4$	Alcohols, ethers	0 to RT	Ketone → alcohol Aldehyde → alcohol
Na(BH$_3$CN)	Alcohols, water, DMSO	0 to RT	Ketone → alcohol Aldehyde → alcohol Alkyl halide → alkane Imine → amine
Na[BH(OAc)$_3$]	Acetic acid	0 to RT	Ketone → alcohol Aldehyde → alcohol Imine → amine
Zn(BH$_4$)$_2$	Ethers	0 to RT	Ketone → alcohol Aldehyde → alcohol
n-Bu$_4$NBH$_4$	Ethers, dichloromethane	0 to RT	Ketone → alcohol Aldehyde → alcohol

(Continued)

Table A6.1 Hydride reducing agents (*Continued*)

Reagent	Typical solvents	Typical temperature (°C)	Functional groups reduced
LiAlH$_4$	Ethers	−78 to reflux	Ester → alcohol Ketone → alcohol Aldehyde → alcohol Alkyl halide → alkane Alkyne → trans-alkene Epoxide → alcohol Imine → amine Amide → amine
Li[(*t*-BuO)$_3$AlH]	Tetrahydrofuran	−78 to RT	Acid chloride → aldehyde Ketone → alcohol Aldehyde → alcohol
Na[AlH$_2$(OCH$_2$CH$_2$OCH$_3$)$_2$] (Red-Al)	Ethers, toluene	−78 to RT	Ester → alcohol Ketone → alcohol Aldehyde → alcohol Alkyl halide → alkane Epoxide → alcohol α, β-Unsaturated ketone → allylic alcohol
B$_2$H$_6$	Tetrahydrofuran, dichloromethane	−78 to RT	Carboxylic acid → alcohol Ketone → alcohol Aldehyde → alcohol Amide → amine
BH$_3$[S(CH$_3$)$_2$]	Tetrahydrofuran, dimethyl sulfide	−78 to RT	Carboxylic acid→ alcohol Ketone → alcohol Aldehyde → alcohol Amide → amine
(Sia)$_2$BH (disiamylborane)	Tetrahydrofuran	−78 to RT	Ketone → alcohol Aldehyde → alcohol Lactone→ lactol
AlH$_3$	Ethers	−78 to RT	Ester → alcohol Ketone → alcohol Aldehyde → alcohol Alkyl halide → alkane Epoxide → alcohol Lactone → cyclic ether

Table A6.1 Hydride reducing agents (*Continued*)

Reagent	Typical solvents	Typical temperature (°C)	Functional groups reduced
i-Bu$_2$AlH (DIBAL)	Ethers, toluene, dichloromethane	−78 to RT	Ester → aldehyde Nitrile → aldehyde Amide → aldehyde Ester → alcohol Ketone → alcohol Aldehyde → alcohol Lactone → lactol α,β-Unsaturated ketone → allylic alcohol Alkyne → cis-alkene

Table A6.2 Single electron transfer reducing agents

Reagent	Typical solvents	Typical temperature (°C)	Functional groups reduced
Li/NH$_3$	Liquid ammonia	−78 to −33	α, β-Unsaturated ketone → ketone Aryl ring → dihydroaryl ring Ketone → alcohol Aldehyde → alcohol
LiC$_{10}$H$_8$ (lithium naphthalenide)	Tetrahydrofuran	−78 to RT	Arylalkylsulfide → alkane Arylalkylsulfone → alkane
Na/Hg (sodium amalgam)	Methanol	0 to RT	Ketone → alcohol Aldehyde → alcohol Arylalkylsulfone → alkane

Table A6.3 Common hydrogenation catalysts

Catalyst	Typical Solvents	Typical H$_2$ Pressure (Bar)	Typical Temperature (°C)	Functional Groups Reduced
Ni (Raney)	Alcohols	1 to 50	5 to 100	Alkene → alkane Ketone → alcohol Aldehyde → alcohol
Pd/C	Alcohols, acetic acid	1 to 10	5 to 100	Aromatic nitro → aromatic amine Alkyl nitro → amine Nitrile → amine Imines → amine Alkyne → alkane Alkene → alkane Alkyl halide → alkane Epoxide → alcohol
Lindlar catalyst (Pd)	Alcohols	1 to 3	5 to 50	Alkyne → cis alkene
Pt/C	Alcohols	1 to 50	5 to 150	Aromatic nitro → aromatic amine Alkene → alkane Ketone → alcohol Aldehyde → alcohol Imines → amine
Rh/C	Alcohols, acetic acid	3 to 50	5 to 150	Aromatic ring → cyclohexane Alkene → alkane Ketone → alcohol Aldehyde → alcohol
Rh/Al$_2$O$_3$	Alcohols	3 to 50	50 to 150	Aromatic ring → cyclohexane

Appendix 7: Common oxidizing reagents

Reagent	Typical solvents	Typical temperature (°C)	Functional groups oxidized
CrO_3/H_2SO_4 (Jones reagent)	Acetone, ether	0 to RT	2° Alcohol → ketone 1° Alcohol → carboxylic acid aldehyde → carboxylic acid
$CrO_3.Py_2$ (Collins reagent)	Dichloromethane, pyridine	0 to RT	2° Alcohol → ketone 1° Alcohol → aldehyde
$CrO_3.Py.HCl$ (PCC)	Dichloromethane, DMF	0 to RT	2° Alcohol → ketone 1° Alcohol → aldehyde
$H_2Cr_2O_7.Py_2$ (PDC)	Dichloromethane, DMF	0 to RT	2° Alcohol → ketone 1° Alcohol → aldehyde 1° Alcohol → carboxylic acid (in DMF)
$KMnO_4$	Water	0 to reflux	2° Alcohol → ketone 1° Alcohol → carboxylic acid Aldehyde → carboxylic acid Alkene → 1,2-diol Cleave alkene → carboxylic acid Sulfide → sulfone
MnO_2	Hydrocarbons, toluene, dichloromethane	0 to reflux	Allylic alcohol → α,β- unsaturated ketone Benzylic alcohol → ketone

(*Continued*)

Reagent	Typical solvents	Typical temperature (°C)	Functional groups oxidized
OsO$_4$/co-oxidant (e.g., NMO)	Acetone/water/ tert-butanol	0 to RT	Alkene → 1,2-diol
OsO$_4$/NaIO$_4$	Dioxane/water	0 to RT	Cleave alkene → aldehyde or ketone
RuO$_4$/co-oxidant	Ethyl acetate/ acetonitrile/ water	0 to RT	Alkene → 1,2-diol Cleave alkene → carboxylic acid
Pr$_4$NRuO$_4$/ co-oxidant (TPAP)	Dichloromethane	0 to RT	2° Alcohol → ketone 1° Alcohol → aldehyde Diol → lactone Sulfide → sulfone
Pb(OAc)$_4$	Acetic acid, acetonitrile	−78 to RT	Cleave alkene → aldehyde or ketone Ketone → α-acetoxy ketone
Ag$_2$CO$_3$/Celite (Fetizon's reagent)	Dichloromethane, hydrocarbons	RT to reflux	2° Alcohol → ketone Diol → lactone
NaClO$_2$/2-methyl-2-butene (Pinnick oxidation)	t-BuOH/water	0 to RT	Aldehyde → carboxylic acid
Pt/O$_2$	Acetone/water	RT to reflux	1° Alcohol → carboxylic acid Diol → lactone
PdCl$_2$/CuCl$_2$/O$_2$ (Wacker oxidation)	Sulfolane/water	RT to reflux	Terminal alkene → methyl ketone
Al(OR)$_3$/acetone (Oppenauer oxidation)	Acetone	RT to reflux	2° Alcohol → ketone
VO(acac)$_2$/ t-BuOOH	Dichloromethane	−20 to RT	Allylic alcohol → epoxide
DMSO/(COCl)$_2$/ Et$_3$N (Swern oxidation)	Dichloromethane	−78 to RT	2° Alcohol → ketone 1° Alcohol → aldehyde Primary diol → dialdehyde
DMSO/DCC/Et$_3$N (Pfitzner–Moffatt oxidation)	Dichloromethane	−40 to RT	2° Alcohol → ketone 1° Alcohol → aldehyde
DMSO/Py.SO$_3$/ Et$_3$N (Parikh–Doering oxidation)	DMSO		2° Alcohol → ketone 1° Alcohol → aldehyde

Reagent	Typical solvents	Typical temperature (°C)	Functional groups oxidized
NCS/(CH$_3$)2S	Toluene	−20 to RT	2° Alcohol → ketone 1° Alcohol → aldehyde
(CH$_3$)$_2$CO$_2$ (dimethyldioxirane)	Acetone	0 to RT	Alkene → epoxide Sulfide → sulfone
m-CPBA	Dichloromethane	−20 to RT	Alkene → epoxide Sulfide → sulfoxide/ sulfone
NaOH/H$_2$O$_2$ (Weitz–Scheffer epoxidation)	Water/ethanol	0 to RT	α, β-Unsaturated ketone → epoxy ketone
O$_3$	Dichloromethane, methanol	−78 to RT	Cleave alkene → aldehyde or ketone

Index